In Sickness and in Health

The Future of Medicine: Added Value & Global Access

IN SICKNESS AND IN HEALTH
THE FUTURE OF MEDICINE: ADDED VALUE & GLOBAL ACCESS

Editor Marleen Wynants

Editorial Board Lieven Annemans, Marc Bogaert, Alain Dupont, Sara Engelen, Joeri Guillaume, Marc Noppen

The editor and the editorial board are not responsible for the content and visions expressed in the articles nor for the correctness of the figures and indicated references. These remain the sole responsibility of the authors themselves.

Cover photo:
'My Smoking Family' by Claire Collison
An image from 'Cradle to Grave' art installation
by Susie Freeman, Dr Liz Lee and David Critchley
The Wellcome Gallery at the British Museum

Book design: Liz Morrison
Printed in Belgium by: Drukkerij Van der Poorten, Kessel-Lo

2009 VUBPRESS Brussels University Press
VUBPRESS is an imprint of ASP nv (Academic and Scientific Publishers nv)
Ravensteingalerij 28
1000 Brussels
Tel: +32 (0)2 289 26 50
Fax: +32 (0)2 289 26 56
info@vubpress.be
www.vubpress.be

ISBN 978 90 5487 549 9
NUR 740 / 860
Legal deposit D/2009/11.161/008

ACKNOWLEDGEMENTS

I want to express my gratitude in the first place to all the authors who made a great effort to finalize their texts while keeping track of their already overloaded agendas. In the second place my thanks go to the people of the accompanying committee and its chairman Alain Dupont for co-thinking and developing this extremely interesting and valuable project. Special thanks go to Marc Bogaert, Lieven Annemans and Joeri Guillaume whose suggestions and critical advice were elementary throughout the whole project's process.

I want to offer a special mention of gratitude to Marc Noppen, CEO of the University Hospital of Vrije Universiteit Brussel, who in a modest way is one of the most supporting forces for anything CROSSTALKS and especially for this project. Furthermore, I want to thank former vice rector Jan Cornelis, Sonja Haesen, Mieke Gijsemans, Monique Peeters, Koen Smets, Christ'l Vereecken, Sandra Baeyens and the whole staff at R&D of Vrije Universiteit Brussel for assisting us where needed.

I want to thank Liz Lee and Susie Freeman for sharing their wonderful border crossing artwork with us, Liz Morrison for her meticulous English editing and beautiful layout and Gert De Nutte for accepting every CROSSTALKS initiative with a broad smile and keeping the presses rolling. I want to thank my daughters Dylan and Ezra for their critical notes on the visual aspects of every new CROSSTALKS book and Luc Steels for his continuing warm encouragement from wherever he is at the moment I start a new CROSSTALKS' chapter. Last but not least, I want to thank Sara Engelen for being my sparring partner in CROSSTALKS' exciting adventures.

Marleen Wynants, March 2009

Accompanying committee

Chairman:
Alain Dupont
Dean of the Faculty of Medicine and Pharmacy, Vrije Universiteit Brussel

Members:
Bernadette Adnet
First Advisor, Social Department, VBO
Lieven Annemans
Health Economist, University of Ghent and Vrije Universiteit Brussel
Marc Bogaert
Professor Emeritus and former Dean of the Faculty of Medicine,
University of Ghent
Julien Brabants
Executive Director Public Affairs, GlaxoSmithKline
Jean-Paul Degaute
President of the Federal Agency of Medicines and Health Products of Belgium
Ri De Ridder
Director-General, RIZIV / INAMI
Stefan Gijssels
Vice President Public Affairs & External Communications,
Janssen Pharmaceutica N.V.
Joeri Guillaume
Study Department Socialist Mutualities
Katrien Kesteloot
Financial Director, University Hospitals, Catholic University of Leuven

Guy Mannaerts
  Former President, University Hospitals, Catholic University of Leuven
Guy Peeters
  General Secretary of the National Association of Socialist Mutualities
Piet Schutyser
  Vice President and Administrator NV AstraZeneca SA,
  CEO AstraZeneca Foundation
Johan Van Calster
  Former Director-General, DG Medicinal Products of the FPS Public Health
Luc Vermeesch
  Vice President Europe, UCB
Kris Westelinck
  Managing Director Pfizer Belgium/Luxembourg
Marleen Wynants
  Operational Director, CROSSTALKS Vrije Universiteit Brussel

Communications officer:
Sara Engelen
  CROSSTALKS Vrije Universiteit Brussel

PROLOGUE

MARLEEN WYNANTS
SARA ENGELEN

In 2005, right after the release of the first CROSSTALKS publication 'How Open is the Future?', CROSSTALKS started developing a project aiming at an open and interdisciplinary dialogue between all stakeholders in health care. Through a series of introductory workshops, we explored the future role, price setting, access to and added value of medicines. An accompanying committee was set up and a three-year project was defined, funded by unconditional grants from Astra Zeneca Foundation, GlaxoSmithKline, Janssen Pharmaceutica, Pfizer, UCB and Vrije Universiteit Brussel. The project started officially in 2006[1] with a first congress and a series of thematic brainstorms and workshops to which all stakeholders were invited.

The ambitious aim was to establish non-mediated and constructive dialogues beyond all the disciplines and organizations involved focusing on the future role and added value of medicines. In parallel, we wanted to construct, gradually, a culture of exchange towards a sustainable and efficient health care. Thus the project set out from a bottom-up and open dialogue between universities, industry, sickness funds, health care providers, physicians, pharmacists, knowledge centers, professional health organizations and policy makers. This cross-disciplinary approach is key to give an incentive to an integrated and a more cost-balanced and qualitative cure and care.

---

1 The complete program, reports and detailed past events can be found on http://crosstalks.vub.ac.be.

Throughout the years we did succeed in establishing a constructive, trust-based open communication between the major stakeholders in health care in Belgium, although we have to continue working on a systematic engagement of all the mutualities, enhancing the dialogue with Wallonia and engaging the representatives of doctors and patients. The results of the constructive interactions manifested themselves in new and various collaborations between the participants and in mind-broadening perspectives on issues like:

- the respective responsibilities of all stakeholders and partners in health care;
- the socio-economic conditions to meet the global need for innovation;
- the common objectives and shared risks in order to render the overall model of health care more transparent and efficient;
- the explicitness of the contrary interests of the different actors in the field and to take these into account during the process of sustainable policy making.

Even more tangible is this present collection of strong and sometimes contesting perspectives, expressed by key players in Belgium, Europe and beyond. 'In Sickness and in Health' presents the particular needs, challenges and aspirations, common aims and perspectives that emerged or were confirmed during the many interactions and workshops.

Setting the agenda for further collaborations, is yet another step. Hence we will start exploring new border-crossings from mid-2009 on, this time zooming in on the urgent need for more transparency between all stakeholders. A transparency that imposes itself every time that concrete action plans are discussed in order to achieve the common goals ahead of us in health care: affordable access to medicines, qualitative cure and care and encouraging innovation and competitiveness.

It is exceptional when scientific experts from competing pharmaceutical companies and academic departments come together not only to share their latest findings on human diseases and treatments, but also to discuss their personal vision of the many complexities in health care. And that is exactly what we want to achieve. We are convinced that there are no shortcuts to true innovation and long-term policy making. In the meantime, we want to offer you these emerging visions as a stepping stone to the next chapter in this constructive CROSSTALKING.

TABLE OF CONTENTS

ACKNOWLEDGEMENTS — 3
PROLOGUE — 7
TABLE OF CONTENTS — 10

PART ONE: UNMET NEEDS

1. **TIKKI PANGESTU** — 15
   The future of health care – A global perspective
2. **PAUL STOFFELS** — 27
   Unmet medical needs: Only cooperation between all partners will lead to solutions
3. **ALAIN DUPONT** — 41
   Access to innovative medicines and orphan drugs
4. **SILVIO GARATTINI** — 57
   The increasing need for independent research
5. **JACQUES DE GRÈVE** — 65
   The cost and added value of personalized medicine
6. **MARC DE VOS & BRIEUC VAN DAMME** — 77
   Breaking the deadlock of budgetary autism: What are the paradigms for future health care organization in Belgium?

PART TWO: SHARED RESPONSIBILITY

7. **ERIK SCHOKKAERT** — 99
   Willingness to pay and solidarity
8. **DANIEL PIPELEERS** — 115
   Notes from a traveler on the road to cell therapy in diabetes
9. **MARLEEN WYNANTS** — 125
   Sharing responsibility in the discovery, development and delivery of medicines: A roundtable session about the actual challenges and potential added value of an industry in transition

10. MAX BAUMANN 143
Why health care systems are not patient-centered
and what this means
11. MARC BOGAERT 155
Physicians and the pharmaceutical industry: A sensitive issue

PART THREE: NEW FRONTIERS

12. GUY PEETERS 167
Common goals to guarantee accessible health care:
A potential win-win situation or just a (bad) dream?
13. ORVILL ADAMS & VERN HICKS 183
Co-payment and future socio-fiscal models
14. THOMAS POGGE & DORIS SCHROEDER 197
Why we need a new approach to pharmaceutical innovation:
A pragmatic answer to a moral question
15. FRANÇOISE MEUNIER 215
Challenges and opportunities of cancer clinical research at
pan-European level
16. SARA ENGELEN & MARC A. MARTI-RENOM 229
Open source drug discovery: The Tropical Disease Initiative

EPILOGUE

SUSIE FREEMAN, DAVID CRITCHLEY & LIZ LEE 245
Cradle to Grave

LIST OF PICTURES 260
INDEX 262

PART ONE
UNMET NEEDS

THE FUTURE OF HEALTH CARE – A GLOBAL PERSPECTIVE

TIKKI PANGESTU

Let me begin with an anecdote. In early 2006 I travelled to Mali in Francophone West Africa where we visited a rural health clinic in Bamba, about 400 km from the capital Bamako. When we spoke to the nurses about their daily work, one of them informed me: "J'arrive à la clinique le matin à huit heures. Il n'y a pas des médicaments à prescrire; à dix heures je vais vendre des légumes au marché". (I arrive at the clinic at eight in the morning. There are no medicines to prescribe; at ten o'clock I go to the market to sell vegetables).

This anecdote illustrates one of the main problems with health care delivery in the developing world: limited or no access to interventions for health improvement. Of course, there are many ways to look at the future of health care – from a national and regional perspective, from a health economics, technology assessment and ethical perspective, and from the perspective of the patient and the consumer. Most of these focus on the cost and efficiency aspects of health care, where the reality of increasing costs due to an ageing population and the high costs of medicines and medical technology are a major cause of concern to governments in the developed world.

Objective

But I am not going to take this perspective. Instead, I will take a different perspective based on my current position at the World Health Organization (WHO) and my own personal training as a scientist. I would like to propose that the future of health care is closely linked to its role in promoting global health in both developed and developing countries. It depends not only on the development of new and improved interventions, but also on ensuring equitable access to such interventions for those who are most in need.

I will try to explain why I think this and what the implications are. I shall also try to frame the presentation in two contexts: one related to the health care delivery system, and the other, with acknowledgement to the mission of CROSSTALKS, to the critical role of research.

Background

By way of introduction, some of the major challenges to global health in the developing world are encapsulated in the Millennium Development Goals (MDGs), a global pact agreed upon by all countries at the UN Millennium Summit in September 2000. Three of these goals (reducing child mortality; improving maternal health; and combating HIV/AIDS, TB, malaria and other diseases) are directly health-related.

In these troubled times, 'health for all' is arguably a key component of equitable and sustainable development and, ultimately, peace and prosperity among all nations and peoples. WHO's Commission on Macroeconomics in Health estimated in 2001 that each 10% improvement in life expectancy at birth is associated with a rise in economic growth of at least 0,3-0,4 percentage points per year. Economists will appreciate the significance of these figures.

Developed countries (e.g. France, Germany, USA, and Japan) spend a considerable percentage of their GDP (Gross Domestic Product) on health care and their populations spend only a relatively small percentage of their disposable income on out-of-pocket payments for health care. They also have sufficient hospital beds, doctors, and nurses and have good vital registration systems for deaths. In contrast, low and middle-income developing countries spend much less on health care, have fewer hospital beds, doctors and nurses and, consequently, much higher infant, maternal and adult mortality rates.

There is clearly a large gap between the developed and developing worlds – only 11% of the total global health budget is spent in low and middle-income countries where 84% of the world's population lives.

Challenges to global health

If we are to narrow this gap between the developed and developing worlds it is important to understand what are the main challenges facing global health:

1. Changing disease epidemiology;
2. Weak health systems;
3. Inequity of access;
4. Gap between research and health care delivery.

Let me elaborate on the first three of these challenges.

Changing disease epidemiology
Developing countries continue to struggle with communicable diseases. The threat of avian influenza continues and, more recently, we have witnessed the appearance of virtually untreatable forms of TB (XDR-TB). At the same time, these countries are now facing a 'double burden' of disease in that chronic diseases (e.g. cancer, cardiovascular disease, diabetes, mental disorders, etc.) are becoming a major public health problem. Of the 35 million deaths every year caused by chronic diseases, 80% now occur in the developing world. The burden of caring for chronically ill patients imposes an additional strain on already fragile health systems and, in addition, the developing world also has to deal with significant mortality and morbidity associated with violence and injuries.

Weak health systems
Eleven million children under the age of five die every year, 90% of them in the developing world. The sad fact, however, is that two-thirds of these deaths (seven million children) can be prevented by existing, safe, effective and cheap interventions (e.g. vitamin A, tetanus toxoid, insecticide-treated bed nets, breast feeding, clean water, etc.). The tragedy is that these interventions are not reaching those who need it most because current delivery strategies are simply not working. It is thus not a surprise that many developing countries are either unlikely or very unlikely to reach MDG 4, reducing under-fives mortality by two-thirds by 2015.

There are two other components of the health system that are in serious trouble in the developing world: financing and human resources. It has been estimated that 140 million people in 44 million households in the developing world face financial catastrophe every year as a result of out-of-pocket payments that exceed 40% of disposable income. As a result, many families will have to forego necessities such as food, clothing and education fees for their children. Some, in fact, do not dare to seek treatment as the drugs, consultations and hospital care can mean bankruptcy for the entire household. Health systems in the developing world also face an acute shortage of health workers. Africa, which carries 25% of the global disease burden, only possesses 3% of the health workforce. In contrast, North America, which carries 10% of the disease load, has 37% of the workforce. It has been estimated that Africa needs an additional one million health workers in the next 10 years if it were to achieve the MDGs.

Inequity of access

Whether one dies of HIV/AIDS, TB, malaria, diarrhea, respiratory diseases or malnutrition, the chances of dying are much greater if one is poor. The poor of the earth bear a disproportionately higher burden of disease than the rich. Women living in South Asia or sub-Saharan Africa bear the highest number of maternal deaths, partly due to the low probability of having skilled attendants present during the birth of their child.

Gap between research and health care delivery

The fourth challenge refers to the gap existing between what could be done and what is actually being done. Reflecting upon the mission of CROSSTALKS, which is to promote dissemination of research and formation of networks between key players, I will elaborate in more detail on this fourth challenge.

However, before analyzing the gaps, let us step back and re-visit the question 'How does research improve health?' According to Julio Frenk, Mexico's Minister of Health, research improves health through three mechanisms:

1. Better interventions;
2. Internalization of knowledge and empowering people to effect behavior change, which results in healthier lifestyles;
3. Informing policy and decision-making.

There have been many studies conducted on the value of research and the benefits or returns to investment that it brings. The latest study from the NIH (National Institutes of Health), focusing on 28 phase III randomized controlled trials in the area of cardiovascular disease and strokes, estimated that the benefit to society is $15,2 billion after 10 years, and that some of the benefits are seen even after only one to two years.

Now to return to the question, what are the components of this gap between 'knowing' and 'doing'?

1. Imperfect R&D process;
2. Weak public trust and confidence;
3. Know-do gap

Imperfect R&D process
Imbalance. Despite the more than $100 billion invested annually in global health R&D (with approximately 45% spent by the pharmaceutical industry), the number of new drugs developed is falling. Also, between 1975-2000, of the 1,393 new drugs developed, only 16 (1,3%) were for tropical infectious diseases prevalent in the developing world. Tellingly, only 10% of these resources are being used to research 90% of the world's health problems (the '10-90' gap). For example, based on R&D funding per DALY (disability-adjusted life years), 16 times more resources are spent on diabetes research (and 10 times more on cardiovascular disease research) than is being spent on malaria research. Another related fact is the development of new antibiotics: despite the fact that antibiotic resistance is a major threat in both developed and developing countries, the major pharmaceutical companies are not investing resources in developing new ones; only three were developed in 2003-2004 as compared to 16 in 1983-1987.

Inefficiency. The research process itself is sometimes inefficient. A total of 64 trials were conducted between 1987-2002 to assess the effectiveness of Aprotinin, a drug used to reduce peri-operative blood loss. However, the evidence that it was effective was actually obtained after the 12th trial in 1992. This effectively means that the additional 52 trials were unnecessary, unethical and wasted valuable resources that could have been used for other research problems.

Biases. There are also biases in the research process itself. Publication bias is well known and, for example, positive results are published more rapidly and research supported by for-profit entities is more likely to yield positive results than that supported by not-for-profit organizations. Biases and gaps also exist in the knowledge base itself in that researchers who work on diseases associated with poverty are much less likely to have their results published in leading international journals.

Limited access. Access to the results of research is often expensive and difficult. Journal subscription rates have been going up steadily in the past 25 years and many are beyond the reach of libraries in the developing world. Internet access is similarly limited and the 'digital divide' is a reality as Africa has only 3% of global Internet use and 95% (of this 3%) is in South Africa.

Neglect of health systems research. Health systems and health services research, arguably critical in promoting global health through strengthening health systems, is a relatively neglected, unglamorous, poorly visible and under-funded field of research. A recent analysis of the major research areas funded by funding agencies in the United Kingdom showed that, on average, health services and health systems research received only 1% of total funding, in stark contrast to biomedical and clinical research. This sad state of affairs is reflected in the scientific output emanating from this area of research: in 2000, only 0,7% of the publications in Medline were on topics related to health services research.

Weak public trust and confidence

Public trust in science, and scientists, has been eroded recently by several instances of scientific fraud, in both basic and clinical research. Major pharmaceutical companies currently have a major trust and image problem with the general public as they have been revealed to conceal safety data from clinical trials. Both Merck and GlaxoSmithKline are facing numerous lawsuits regarding the safety of Vioxx and Paxil respectively. The former is a pain-reliever linked to increased risk of heart disease, and the latter an antidepressant associated with increased risk of suicidal and violent behavior in adolescents. The pharmaceutical industry has also been accused of 'disease mongering', of inventing illnesses and turning everyone into patients for the sake of profits.

The safety of clinical trials has also come under close scrutiny following the disastrous phase I clinical trial of TGN1412 in the United Kingdom recently. Injection of small doses of this immunomodulatory drug, a humanized monoclonal antibody acting as a superagonist to the CD28 receptor of T cells, resulted in a severe adverse reaction and hospitalization of six volunteers, at least four of whom suffered major organ failure, severe depletion of regulatory T cells and are likely to suffer immune system-related problems for life. More importantly, many clinical trials are likely to be conducted in the developing world in the future and concerns have been expressed with regards to ethical and regulatory aspects. Russia and India, for example, have seen a 200% increase in the number of trials in the past five years.

[21]

The public is also dismayed by the element of uncertainty associated with research and science as they encounter it in their daily lives. There is a distinct impression that *"if even the experts can't agree, what hope is there for the rest of us"?* The benefits of statins in preventing heart disease and hormone replacement therapy for post-menopausal women are just two examples where leading researchers have given conflicting opinions.

Poor link between evidence and decision-making

To link research with action is a complex and difficult challenge. As once suggested, *"politicians don't see the light; they feel the heat"*. Researchers and policy and decision-makers often have different mind sets, values, time lines and priorities. The following summarizes the conflict between policy makers (PM) and researchers (R): PM deals with complex policy problems, R focuses on the simplification of problems; PM has to come up with focused solutions, R is interested in related but separate issues; PM wants to reduce uncertainties, R is interested in finding the truth; PM emphasizes speed, R wants time to think; PM wants to control and delay, R wants to publish or perish; PM thinks of manipulation, R thinks of explanation; PM wants feasible and pragmatic solutions, R is interested in thoughtful deliberations.

Evidence-informed action is arguably even more important for developing countries who are faced with limited resources and competing priorities. As stated by Dr Hassan Mshinda (Ifakara Centre, Tanzania): *"If you are poor, actually you need more evidence before you invest, rather than when you are rich"*.

What now? Actions to overcome challenges

Identifying the gaps between research and health care delivery is only the first step. What actions can be taken to address these gaps? Here I give some selected examples:

Better define research priorities

An example of a recent effort to better define research priorities for the developing world is the DCP-2 initiative (Disease Control Priorities) sponsored by WHO, the World Bank and the Bill & Melinda Gates Foundation. The DCP-2 initiative identified three major research priorities: (1) new and better drugs, vaccines and diagnostics; (2) national and global surveillance; (3) strengthen health systems.

Novel ways of developing interventions for unprofitable diseases

Given some of the limitations to current strategies for developing new interventions, especially for diseases afflicting the developing world, novel and innovative alternative strategies should be explored. This may include Public-Private Partnerships (PPPs), such as the Institute for One World Health and the Medicines for Malaria Venture (MMV), partnerships between governments in the developed and developing world (EDCTP – European Developing Countries Clinical Trials Partnership), partnerships between industry and government (e.g. the Novartis Institute for Tropical Diseases in Singapore, GSK's Tres Cantos research unit in Spain), and the open source approach to drug discovery, development and manufacture (e.g. the Tropical Diseases Initiative – see chapter 17 of this book – BioForge). With regards to the PPPs, it has been estimated that they now account for 75% of all R&D projects in neglected diseases.

Improved access to published research results

Open access peer-reviewed journals like PLoS Medicine would clearly help to improve access to the results of research. Many major research funding agencies in the United Kingdom and the NIH in the USA, for example, now require results of the research they support to be posted on such open access sites. Also, the WHO brokered an agreement with 90 major publishers to provide free, online, full-text access to more than 3,000 journals to researchers in 100 developing countries.

Promoting transparency and accountability
One initiative which aims to strengthen public trust by promoting transparency and accountability in clinical research is WHO's initiative to establish an International Clinical Trials Registry Platform. This voluntary platform encourages the registration of all interventional clinical trials (including exploratory phase I trials) at the time of commencement and before enrolment of the first patient or volunteer. It requests a minimum registration data set of 20 items to be fully registered, without provision for any delayed disclosure on any of the items.

Evidence to policy: Learning from what works
In a philosophy captured as 'The Triangle that Moves the Mountain', Professor Prawase Wasi from Thailand articulated that change must be driven not only by knowledge but must also include the elements of social mobilization and political commitment. When these three elements are in place, evidence has effectively shaped policy.

[23]

One example comes from the Tanzania Essential Health Interventions Project (TEHIP) where community-based participatory research, designed together with district health officers, documented the true burden of disease in the Rufiji district of the country. Following the outcomes of the research, the distribution of the district health budget was adjusted to better reflect the true disease burden. Since the commencement of the project in 1997, the district has witnessed a 52% decline in the under-fives infant mortality rate which is a truly impressive achievement.

The second example is at global level: scientific evidence on the harmful effects of tobacco was used by WHO to develop the first ever Framework Convention on Tobacco Control (FCTC). The treaty came into force in February 2005 and is a legally binding international instrument to commit countries to curb the use and sale of tobacco products within their national borders, e.g. by banning tobacco advertisements on television, prohibiting sale to minors and requiring prominent labeling of tobacco products. The FCTC will hopefully play an important role in reducing the five million deaths per year related to tobacco use in both developed and developing countries.

Summary and conclusions

Health care in the future must be closely linked to promoting global health for all peoples, in both developing and developed countries, in an equitable and sustainable manner. 'Health for all' is now a common concern for us all and is a major driver of global peace, development and security. I have outlined the challenges, gaps and possible actions, but whether or not we succeed in reaching our common objective of overcoming inequity and imbalances will depend on all key stakeholders working together in innovative and effective partnerships based on trust, transparency and open and constructive dialogue.

SELECTED READING

Macroeconomics and Health: Investing in Health for Economic Development. Report of the Commission on Macroeconomics and Health. World Health Organization, Geneva, 2001.

World Report on Knowledge for Better Health. World Health Organization, Geneva, 2004.

D. de Savigny, H. Kasale, C. Mbuya, G. Reid. *Fixing Health Systems.* IDRC, Ottawa, 2004.

Preventing Chronic Diseases: A Vital Investment. World Health Organization, Geneva, 2005.

R. Moynihan, A. Cassels. *Selling Sickness: How the World's Biggest Pharmaceutical Companies are Turning All of Us Into Patients.* Nation Books, New York, NY, 2005.

Working Together for Health. The World Health Report 2006. World Health Organization, Geneva.

Priorities in Health. The World Bank, Washington, DC, 2006.

I. Evans, H. Thornton, I. Chalmers. *Testing Treatments: Better Research for Better Healthcare.* British Library, London, 2006.

BIO

Tikki Pangestu is presently Director of the Research Policy & Cooperation Department of the World Health Organization in Geneva. He holds a BSc (Hons) degree in biochemistry and a PhD in immunology from the Australian National University in Canberra, Australia. He was previously Professor of Biomedical Sciences, Institute of Postgraduate Studies & Research, University of Malaya, Kuala Lumpur, Malaysia, and editor-in-chief and publisher of the Asia Pacific Journal of Molecular Biology & Biotechnology. His professional accreditations include FRCPath (Fellow, Royal College of Pathologists, United Kingdom); FIBiol (Fellow, Institute of Biology, United Kingdom); FAAM (Fellow, American Academy of Microbiology, USA); FAMM (Fellow, Academy of Medicine of Malaysia); Fellow, Academy of Science of the Developing World; Member, International Molecular Biology Network (IMBN). His research and academic interests include the prevention and control of infectious diseases, development of research capabilities in developing countries, assessment of health research system performance, evidence to policy linkages, impact and application of modern biotechnology on developing economies.

UNMET MEDICAL NEEDS: ONLY COOPERATION BETWEEN ALL PARTNERS WILL LEAD TO SOLUTIONS[1]

PAUL STOFFELS

Despite all the progress made over the past 50 years, the world's population is today still faced with massive health problems and a need for therapeutic innovation. We have arrived at a turning point in the way we approach innovation, a moment where the balance between investment risks and rewards is in jeopardy, a moment where health care budgets are under heavy pressure due to amongst others ageing, and therefore also a moment for all parties involved, be they national, regional or international, private or public, to reflect on their role in bringing health to the patient.

Many parties actively support this debate. And today, we have thankfully grown beyond the time when it was marked by prejudice and controversy. Destructive forces have tainted the dialogue in the past. It is now time to look to the future. In order to arrive at a solution to the world's unsolved medical needs, we have to ask ourselves certain questions:

1. What are the medical needs?
2. What is already happening today and by whom?
3. Who is paying for it all?
4. What are the risks of the current model?
5. What do we have to do next?

I shall try to formulate an answer to all these questions in the hope of laying some fertile ground for a fruitful discussion later on.

---

1   Speech given by Dr Paul Stoffels, Company Group Chairman Johnson & Johnson Pharmaceutical Research & Development, on the occasion of the CROSSTALKS conference in October 2007.

What are the medical needs?

The World Health Organization (WHO) estimates the number of known diseases at around 30,000. Today, we know of treatments for around 6,000 diseases. Many of these as yet untreatable diseases are very rare hereditary conditions, for which a solution is far from within sight. However, even amongst the 6,000 diseases for which solutions already exist, there are still many people for whom the available treatments do not work or do not work to a sufficient extent in terms of effectiveness and side effects. There is always room for improvement.

Our own company, Johnson & Johnson, has made the search for solutions to these medical needs the most important objective in its research strategy. A few examples are dementia, focusing predominantly on Alzheimer's disease, hospital bacteria, tuberculosis, and HIV/AIDS. For some of these diseases, treatments already exist, however they are far from satisfactory. We are also active in cancer research. The pharmaceutical industry has made gigantic contributions over the past decades to the treatment of cancer. Today, breast cancer patients have a 90% chance of survival, calculated five years after diagnosis, melanoma patients a 92% survival rate, prostate cancer patients 100% and bladder cancer patients 82%. Of course, these figures presuppose timely diagnosis. They represent gigantic breakthroughs in just a few decades time. On the other hand, the survival rates are only 45% for patients with cancer of the fallopian tubes, 16% for lung cancer patients and 5% for pancreas cancer patients [2]. There is still a lot of work to do in the domain of pharmaceutical research.

And that research is critical to individual patients, for whom a few additional, quality years of life can be precious. It is also essential for our society. A study published in October 2007 by the American Milken Institute calculated that for the US alone, the cost of unsolved diseases would amount to a trillion dollars by 2023 ($1,000,000,000,000). Just for the United States [3].

---

2   Surveillance, Epidemiology, and End Results Program, 1975-2003, Division of Cancer Control and Population Sciences, National Cancer Institute, 2006.
3   *"An Unhealthy America: The Economic Burden of Chronic Disease"*, Milken Institute, October 2007.

The consequence of an ageing population on health care is another example of the importance of innovation. If we were to succeed in postponing the cognitive decay in Alzheimer patients by three and a half years, we would decrease the number of patients with the disease by one third[4]. If we could postpone it by five years, we would even halve the patient population.

For developing countries, the overall cost of disease burden is even greater, because public health is one of the basic requirements for emerging from their current painful predicament.

The second question is: Who is doing what to find a solution to this situation?

Naturally, there is a degree of overlap between some of the actors and a lot of cooperation as well, but overall, this is where we stand.

The *academic world* carries out fundamental research in biomedicine and pharmacology, and it also carries out research into disease mechanisms. On the basis of this information, the *pharmaceutical and biotechnology industries* start searching for treatments.

The *health care sector* (doctors, pharmacists, nursing staff,...) ensure that patients receive their treatment.

The *governments* are responsible for the provision of health care in the broadest sense. They are responsible for prevention, for the organization of health care, its accessibility, the funding of reimbursement of products and services, etc.

The *health insurance funds and private insurers* jointly ensure the management of health insurance and provide mechanisms of solidarity between their members.

---

4   Rozzini R, Ferrucci L, Losonczy K, Havlik RJ, Guralnik JM. *Protective effect of chronic NSAID use on cognitive decline in older persons.* J Am Geriatr Soc 44: 1025-9, 1996.

The third question is: Who finances these new treatments?

If we look at what governments spend on health care research, then the United States is by far the largest financer. The American government spends more than $47 billion annually on health care research in the broadest sense. This research is epidemiological, academic and clinical in nature and also includes research into socio-demographic factors, environmental effects on health as well as direct financing of therapeutic research. The European Union spends approximately $8 billion per year, that is the European Commission and all the individual Member States combined[5].

In other words, that is roughly $55 billion spent by the richest countries in the world on health research in the broadest sense. The global pharmaceutical and biotech industry spends a similar amount, namely $59 billion, on research into new pharmaceutical treatments[6]. That is without taking into account research for medical equipment, diagnostics, medical imaging and the like.

The pharmaceutical and biotech industry can do this because it works within a market model. Shareholders invest their resources in a company, which devotes those resources to research, which will hopefully yield products with a high added value. Once on the market, governments will then reimburse them. Thanks to the market that is thereby created, the company behind the innovation acquires the means to compensate its shareholders.

It is very important for all parties to understand this mechanism well. The shareholders, that's you and me, but also the major investment funds and pension funds. They want to see returns for their investment. If that return never comes, then they will go elsewhere with their money, to companies that offer greater returns or sectors with less inherent risk. On the other hand, the pharmaceutical sector cannot be blind to the fact that the reimbursement of our medicines is absolutely essential to generating a market for us.

---

5   These calculations have been made on the basis of various sources: NIH, OECD, and European Commission.
6   Source: PhRrma, EFPIA 2006.

Without government finance via reimbursements, we would not be able to compensate our shareholders. Therefore, without compensation for the risk of shareholders, there can be no innovation. Therefore, without reimbursement by the governments, there can be no innovation.

And the pharmaceutical industry invests gigantic amounts in R&D. Today, the world market for all medicines totals around $600 billion. The R&D investments made by pharmaceutical companies are therefore around 10%. Of this $59 billion, the lion's share is spent by the world's top 20 pharmaceutical companies[7]. Twenty companies out of the estimated 400,000 pharmaceutical companies in the world spend a combined $54 billion on R&D into new medicines. On average, these companies invest around 15% of their turnover in R&D. There is not a single company from any other sector that takes that kind of risk.

Last year, our company spent more than $5 billion on research and development into new medicines against a medicine-driven turnover of $23 billion. In other words, some 21% of the income that our pharmaceutical business generates flows back into research and development for new innovations.

Our research domains correspond to a large extent with what WHO defines as priority domains for pharmaceutical research: antibiotics for resistant bacteria, cardiovascular diseases, HIV/AIDS, tuberculosis, cancer, Alzheimer's, depression, arthritis, diabetes. The reason for this is simple. There are already treatments for simple diseases. Those who come out with a treatment in these domains do not have an immediate market and, consequently, are unable to generate sales or returns for their shareholders. A 'me too' strategy is, in this sense, a commercially risky option. Only strong therapeutic improvements are able to create a market.

It is also often thought that public funding is the main source of pharmaceutical research. This is contradicted by the reality of the situation. Companies are often only involved in research consortia for the European Framework Programmes at the request of the academic world. The public funds we receive in Europe have amounted to around €1 million per year over the last 10 years. That is equivalent to 0,06% of our total expenditure on research into

---

7   Source: European Commission: *"European Innovation Scoreboard"*, 2007.

new medicines. Or, put differently, 99,94% of all our R&D investments are financed by profits from the sale of our current medicines.

What are the risks with the current model?

Shareholders will only continue to invest if they are rewarded for the financial risk they take. They are not going to take risks if they can get the same financial return with lower risks. In such cases, they would prefer to finance for instance the food industry or the metal industry: less risk, certain return. And do not underestimate this factor. In 1990, the pharmaceutical industry invested $15 billion in R&D; today the figure is four times as much. In that period, the number of medicines that reached the stage of clinical research has also drastically increased. In 1996, 380 Investigational New Drugs (IND) were approved for human testing. In 2006, the figure was 700. In 1996, there were 56 new molecular entities, i.e. medicines with a novel active ingredient, approved for registration on the market. In 2006, however, there were only 22 [8].

In table form, this gives:

	1996	2006
Pharma industry investments	$26 billion	$59 billion
INDs approved	380	700
FDA approvals	56	22
Annual cost price/approval	$0,5 billion/NME	$2,6 billion/NME
Annual success rate	14,7%	3%

This also means that in 1996, we spent $0,5 billion per approved compound. Today, that is $2,6 billion, or a decrease in the success rate from 14% to 3%. I am aware that this comparison of investments and results in the same year is somewhat misleading. They refer to different compounds, after all. However, for a company, this kind of comparison is essential. Decisions must be made on the basis of today's results and success rates. And the trend and the scale give the right indications.

---

8   Source: US Food & Drug Administration (Center for Drug Evaluation & Research), PhRma.

If we look at this from the point of view of the life cycle of a single medicine under research, this means that after a period of eight to 10 years of research, we still have only a 66% chance that the product will ever see the market. In our sector, it is all or nothing. And in previous years, it has been 'nothing' for many products from many top companies. If during the construction of a new Airbus, you come across problems with the wing or the landing gear, you adapt the design. At no point during the innovation trajectory of a new airplane or a new car can the entire project come under threat. Our sector is different.

And the risks are mounting, while the potential rewards are declining. Shortsighted critics may well look on with satisfaction, however, the consequences for the patient are potentially enormous. Why are the risks increasing? For the companies, all simple therapies have already been found. Today we are focused on complex diseases such as cancer or diseases of the central nervous system. The investments required are gigantic, not only for basic research but also in the development phase, because more clinical research is required than before, which is conducted over longer periods of time, requiring more patients, more elements to analyze and more countries to be involved.

On the other hand, the demands governments place on medicine safety are ever increasing. And even though many of the additional requirements are justified, each additional demand has an impact on the innovation trajectory, because it influences the economic viability of the research.

In addition, there are the market conditions, which are extremely important in terms of compensation for innovation. It cannot be sufficiently emphasized that the research into tomorrow's medicines is financed by today's sales. In other words, our existing products finance tomorrow's innovation. And today's market is under enormous pressure because health insurance budgets are of course also limited. The population over the age of 60 years also consumes more than 60% of all health care. And that group is only getting bigger. People live longer thanks to the pharmaceutical innovations of the past, but often as chronic patients, which means they have to take medication on a daily basis. This evolution from acute infectious diseases to chronic diseases represents a significant burden on health care budgets. Governments are therefore forced to carefully control budgets, with the consequence that innovation is not always sufficiently rewarded. Governments delay decisions regarding reimbursements and negotiate over the price. Restrictions are imposed on reimbursed usage, reducing the volumes of innovative drugs. In the past, if a company launched a very effective medicine with few side effects, it quickly became the reference in the marketplace. Today, this is no longer the case. Because innovation costs more and more money, most new medicines are also considerably more expensive. As a result, governments limit the use of innovative medicines to those cases where there is no effective therapy yet. In such cases, you might get a good price, but the volume can be effectively reduced to one tenth of that necessary in order to be profitable.

So what could the consequences be of this?

Because the profits of the pharmaceutical industry are under severe pressure, companies tend to take less risk. The limited funds available will go into therapies where the chances of commercial success are greatest. Whereas in the past, all medicines that were effective and safe could count on reaching the market, this is no longer the case today. Today, dozens of medicines will never see the light of day, which would have reached the market ten years ago[9]. And this is not just because the authorities do not admit them. Research demonstrates that more than 30% of the candidate medicines do not reach the market because they are not perceived as feasible by the industry[10]. This means that the industry thinks it will not be able to claim enough added value for these medicines in order to recuperate the cost of the research in the price. Or, as is also the case, that despite the high therapeutic value of the medicine, the development costs are such that the subsequent price can nonetheless not be justified. Medicines that cost several tens of thousands of dollars per patient per year are no longer out of the ordinary. What should a company do at that point? Continue with the research, afraid that it will never be profitable? Or stop the research, knowing that a potential therapy is being abandoned? Who is the company to decide all this by itself? Should the authorities not be involved in this decision-making process sooner?

---

9   Put differently, there are many efficacious and safe drugs on the market today that would no longer pass the current requirements for approval.
10 Adapted from: Kola and Landis, *Nature Review Drug Discovery*, 2004 (3):711-715.

The only solution for the industry is to follow a dual trajectory.

First, the industry must apply a model with more innovative working methods. After all, we start research on the basis of today's regulations. And we know that those regulations are getting stricter and stricter. But at the same time, the capacity for innovation during the development of a medicine is declining: adjustments are made here and there in order, for example, to reduce the toxicity and increase absorption, etc. The probability is therefore real that a medicine will no longer meet the requirements by the time it reaches the market, however innovative its initial concept may be. The only way to avoid the deadlock is to focus on the most innovative products, those which will continue to meet the highest needs despite the increase in regulation. This trajectory is more risky and the potential rewards should be proportional to it.

The second trajectory is to carry out incremental innovation with existing medicines. This could be through the development of new ways of administering drugs in order to achieve greater confidence in use, more comfort and easier administering, etc. These are often, but not always, lower risk investments that do generate returns. There are many – deliberate or not – misunderstandings about incremental innovation.

Many incremental innovations offer high therapeutic added value to patients. A breakthrough medicine may open new avenues for treatment, but the 'first in class' is not necessarily the 'best in class'. Improvements can be made with significant and relevant improvement for patients and public health. A good example is our own treatment for schizophrenia, which was a breakthrough discovery fifty years ago, which underwent many changes and improvements over the decades to reach its current therapeutic potential. The latest development, the injectable form administered every two weeks, offers a major improvement over the daily oral form, increasing treatment compliance and decreasing relapse, the most cumbersome and costly factor in the treatment of schizophrenia patients. This is an incremental innovation, and one like many others that requires fully-fledged clinical studies before being approved by the authorities. The investment risks are less for the company, but the development costs are as high as with breakthrough medicines.

But what do we do then with medicines for diseases that are not profitable, which do not answer to the mechanisms of the market? For these medicines, more direct cooperation is required. As we have said earlier, public health is a responsibility of the government and not of the pharmaceutical industry. Organizing public health in a country is the task of the public authorities, and this implies that they must ensure that there is sufficient research into medicines for which there is no commercial market.

Looking at the figures, we see that today, 70% of research into neglected diseases is financed within Public-Private Partnerships (PPPs), with 50% of the investments coming from multinationals. Today the largest financiers are to be found in foundations such as that of Bill & Melinda Gates, now also partly financed by Warren Buffett. However, the resources devoted to neglected diseases are but a fraction of the total. The Global Fund today has a budget of $4,6 billion for a period of two years. This is a considerable increase in resources, which we cannot but applaud, however it is still far from what is required.

So what should we do?

First of all, the industry has to concentrate more on breakthrough medicines. This calls for massive amounts of money that can only be financed through the sale of today's medicines.

Second, the authorities must evaluate registration applications much faster than is currently the case. Taking one year to evaluate a file is too long. The waiting period should be cut in half.

Third, the authorities should approve and reimburse medicines much faster, and only afterwards begin large-scale analysis with strict controls and follow-up. This could imply that phase three clinical research would be the first market phase. In this phase, the industry could already start recuperating its investments. This means that on the basis of good phase two studies, a conditional market approval would be granted. The (subsidized) medicine would then be released onto the market while the company commits to carrying out large-scale phase three research in strictly controlled conditions. Once that is completed, the conditions can be re-evaluated on the basis of the results. This process already exists as the fast track approvals of the

FDA, but they are currently restricted to life-saving drugs, such as some of our own HIV/AIDS compounds. Expanding this approach to a broader range of drugs could dramatically reduce investment risks and costs, while at the same time giving earlier access to patients.

Fourth, innovation must be rewarded through prices and reimbursements. Today, the conditions placed on innovative medicines are too restrictive. The reimbursing authorities also have to be aware of the fact that the sale of medicines today, and of incremental innovation, pays for research into the medicines of tomorrow. They should therefore accept their share of the risk. Offering insufficient compensation for those medicines is potentially catastrophic for innovation in the future.

Fifth, the authorities must significantly increase their investment in neglected diseases. This is the task of the government, and industry can contribute with its expertise in the discovery and development of new medicines.

Sixth, the world's governments must create a harmonized regulatory system for clinical research. A lot of money is invested today because the demands in Europe, the United States and Asia are often different. The development of medicines should be to the benefit of patients all over the world. And their needs are the same, regardless of where they live.

Finally, we must join forces in finding ways of solving the problem of access to medicines in developing countries. And I assure you: this is not a problem that the industry can solve alone. Even the cheapest medicines are not getting to the people who need them: the patients. It is the task of the national and international authorities to safeguard legal certainty, distribution and follow-up. We, as industry, can and should of course contribute. I will give you a few examples of what we currently do.

In the area of research, we currently have ongoing projects for a microbicide that we have discovered, a potential medicine that protects against the transmission of the AIDS virus during sexual intercourse. This medicine is being further developed within the International Project for Microbicides (IPM). We have donated the medicine to this consortium free of royalties for use in developing countries.

Our very promising molecule for tuberculosis, the first major innovation in forty years, will also only be suitably developed thanks to co-financing by international consortiums.

However, the problems are not limited to the discovery of new treatments. It is not because a product is launched, that it reaches the patient. And price is not the only issue keeping therapies away from patients. Cheap, yet still very necessary medicines are not reaching the patient, and new, expensive and life-saving drugs are too expensive for people in developing countries. For instance, around 400 million children in the world suffer from intestinal worm infections. Such parasites can consume more than half the child's daily requirements of protein. The consequences for children are serious: the disease interferes with cognitive and physical development, and increases the child's susceptibility to other diseases. Our very effective and very cheap medicine for this disease, however, is not getting through to the patient. Which is why we have now initiated a worldwide program to eradicate these infections. After several pilot projects with NGOs, we have begun to initiate the program on a large scale. This year, we will be treating some 35 million children. Over the coming years, we will expand our efforts country by country. This only works, however, in well-supported projects that offer basic training in hygiene and prevention, which we will also co-fund.

We have an anti-HIV/AIDS medicine that we have released in Belgium and the rest of the world, including Africa. In order to reduce the price in Africa, we have called on a manufacturer of generic medicines in South Africa, and we are currently investigating partnerships in Asia. The local partner will also be responsible for distribution. The price there is naturally a lot lower than it is here. And in order to counter re-importation, the pills look different.

Finally, I would like to briefly repeat the most important points of my argumentation. The innovative pharmaceutical industry is the most important investor and source of new and vitally necessary medicines. If there is insufficient reward to offset those investments, then the long-term consequences for the patient will be dramatic. In the interests of patients, we must be able to address today's problems. Let us therefore look for solutions that are acceptable for everyone. We cannot do it alone. Nor can the authorities or the NGOs. Dialogue and cooperation are a must.

BIO

Paul Stoffels studied Medicine at the University of Diepenbeek and the University of Antwerp in Belgium and Infectious Diseases and Tropical Medicine at the Institute of Tropical Medicine in Antwerp, Belgium. He worked for four years as a Physician C Researcher in the field of HIV/AIDS in Africa. In 1991 he became the Head of Development for the HIV compounds at the Janssen Research Foundation in Beerse, Belgium and in 1993 he was promoted to Director Clinical Research and Development for Infectious Diseases and Dermatology. He was the co-founder of two Belgian biotech companies, Tibotec and Virco. In 1997, Paul left Janssen Research Foundation to become the CEO of Virco. Later he was appointed President of Tibotec. He led the development of Tibotec from a technology-based research company into an integrated pharmaceutical R&D company focusing on the discovery and development of new drugs and diagnostics for HIV/AIDS and other infectious diseases. Paul has been a co-founder of a number of other biotech companies in Europe. In April 2002, Johnson & Johnson acquired Tibotec-Virco. Currently, Paul is Company Group Chairman of Pharmaceutical Research & Development at Johnson & Johnson and Chairman of Tibotec. He has special interests in developing solutions for health care problems in the developing world and for optimizing patient management in the developed world through the integration of therapeutics, diagnostics and information systems.

ACCESS TO INNOVATIVE MEDICINES AND ORPHAN DRUGS

ALAIN DUPONT

Submissions for Market Authorization Application (MAA) for medicines for human use are regulated within the European Union (EU) with relevant procedures at Community level in which the value of medicines is considered in terms of efficacy, safety and pharmaceutical quality[1]. *Regarding reimbursement submission however, individual Member States are competent*[2].

In recent years, various Member States (e.g. Belgium, the Netherlands) evaluated the relative *therapeutic value* of medicines using additional criteria such as effectiveness, applicability and convenience to patient or physician[3,4]. In France, weights are given depending on the severity[5] of the disease and the added value ranges from none to substantial on a five-point scale. The introduction of *relative efficiency* (cost-effectiveness) evaluations in reimbursement procedures is different across Member States: the most extensive experience being held by the UK where NICE[6] combines the evaluation of the clinical and economical evidence in technology appraisals.

Major differences in procedure between submissions for market authorization or reimbursement relate to procedural level (mainly centralized versus per Member State), criteria for assessment (efficacy and safety versus additional elements), hypothesis (individual drug benefit/risk ratio versus (added) value compared to therapeutic alternatives) and comparator (mainly placebo versus active comparator).

According to the *Transparency Directive* (TD)[2], pricing and reimbursement decisions must be taken in a *transparent, objective and verifiable way* in respect of *strict time lines* (maximum of 180 days from submission to decision). In the absence of a common procedure, one may expect substantial diversity across Member States in reimbursement procedures and decisions and, as a consequence, in the access to medicines.

Evidence based reimbursement of new medicines in Belgium

In order to be compliant with the TD, various Member States adapted their reimbursement process by integrating Evidence Based Medicine (EBM) principles in the evaluation process. Since 1 January 2002, the reimbursement procedure in Belgium was modified to reduce time lines and to implement EBM principles to enhance transparency and use of objective criteria. The Belgian reimbursement decision by the Minister of Social Affairs is preceded by the evaluation by an advisory committee, the Commission on Reimbursement of Medicines (CRM) of the relative therapeutic value of a medicinal product, based on five defined criteria: efficacy, safety, effectiveness, applicability and convenience.

For the assessment of *efficacy and safety*, all ICH GCP procedures relating to clinical research are applicable if adopted by EMEA[7,8]. The Randomized Controlled Trial (RCT) remains the golden standard for assessing the causal relationship between treatment and outcome, minimizing the possibility of bias[9]. Only peer reviewed full papers or ICH End of Study Reports are accepted: abstracts are refused because they are considered unreliable as has been confirmed by Gøtzsche[10].

For the assessment of *applicability* (contra-indications, special precautions), the official information contained in the Summary of Product Characteristics is used.

Effectiveness and *ease of use* for patient or physician relate to daily practice: effectiveness – the performance of a medicine in general use – cannot be predicted from efficacy as captured in RCTs[9] mainly because the results from a strict RCT study population may suffer from poor translation to a broad daily practice population. In this respect, observational studies on the daily use of medicines may provide *complementary* evidence[11] to that coming from RCTs. The need for both, RCTs and observational studies, has recently been emphasized in order to improve the quality of care[12].
Because there is actually no European GCP counterpart for observational studies, a specific procedure on the design, analysis and reporting of obser-

vational studies for reimbursement purposes has been implemented. The procedure is based on the International Society for Pharmacoepidemiology guideline [13].

The *assessment* of the performance of the submitted medicine versus its alternatives on each of these five criteria results in an *appraisal* by the CRM advisory committee of the relative – added or similar – therapeutic value. The process of assessment and appraisal of the therapeutic value is rather similar to the reimbursement process adopted by other Member States like the UK [6], France [4] or the Netherlands [14] even if differences may exist in criteria used and the composition and role of the advisory committee to the Minister. The appraisal is based on individual weighting of each of the five criteria.

The reimbursement process adopted by the Belgian competent authority yields two distinct and sequential phases:

Phase 1: An EBM evaluation of the therapeutic value of medicinal products based on five *defined* criteria: efficacy, safety, effectiveness, applicability and convenience; the end-points considered must relate to mortality, morbidity or quality of life. The result of the evaluation is binary: either *added* (Class 1) or *similar* (Class 2) therapeutic value. Generic drugs and copies are Class 3 submissions.
Phase 2: A CRM proposal on reimbursement modalities leading to a binary decision to reimburse the medicinal product: the Minister takes this decision based on the advice of a two-thirds majority of the reimbursement commission. The Minister may only deviate from the advice of the commission for specified social or budgetary reasons. In the case of a positive decision, the reimbursement basis and reimbursement modalities are specified. In the case of a negative decision, the decision must be justified. The applicant may appeal in Court against the decision within 60 days.

These sequential steps in the procedure are crucial to disentangle the decision on the *relative therapeutic value* of a medicine from any reimbursement decision in which financial elements (budget impact, cost-effectiveness) are introduced: this two-phase procedure enhances a better understanding of the weights given to scientific versus financial elements in the final decision.

In order to respect strict time lines, the procedure states that if a Ministerial decision is not made within 180 days the applicant obtains unconditional approval of his request. The decision then enters into application on the first day of the month following publication in the 'Official Journal'.

EBM and its impact on the decision outcome:
Analysis of the reimbursement decisions

Using the administrative database from the National Insurance Agency, all files submitted from 1 January 2002 and for which the procedure ended by 31 December 2004 were analyzed. During that period, 1,642 files for reimbursement were submitted. By the end of 2004, there were 1,285 dossiers for which a decision on reimbursement was taken: 67 Class 1 and 92 Class 2 New Molecular Entities (NME) were handled during that period: combining both classes yields 159 requests for new medicines or 12,4% of all handled submissions, with a drop below 10% occurring in the second semester of 2004. Within the combined group of new molecular entities, 42% (n=67) were a request for Class 1 or added value medicine, representing 5,2% of all submissions. Overall, more than one-third of the handled submissions concerned generic compounds reaching 40% in the second half of 2004. An increase over time in the number of requests for new indications for already available drugs was also noted, reaching 24% in the second semester of 2004. The median number of submissions per semester was 251 in the period 2002-2004.

The added value claim was approved in 47,8% of the Class 1 submissions, which means that 20,1% of submissions for NME were considered Class 1 products or on average 11 products per year. The results are similar to the percentage of biotech medicines with superior therapeutic value, which has been estimated at 20%[15].

In France, the number of NME submissions for which a quantitative rating of the added therapeutic value was attributed in 2005[16], was 108: added value was considered major or important for nine products, which is a similar number to the Belgian average number of Class 1 products. Clearly, the vast majority of NMEs do not offer added therapeutic value.

The drop below 10% of NMEs occurring in the second half of 2004 may be related to the fall-off in R&D productivity and a decline in the number of new drugs coming to market as described in the study report on behalf of the European Commission on 'Innovation in the pharmaceutical sector'[17]: this report states that the number of approved new active substances by EU centralized application or Mutual Recognition has fallen from 54 in 2001 to 42 in 2002 and 31 in 2003. A similar decrease in the number of new drug applications submitted to the FDA was also described[18]. Due to the time lag between MAA approval and the local reimbursement decision, this may well explain the drop by the end of 2004.

The percentage of positive decisions for reimbursement was dependent on the type of submission with the lowest percentages for Class 2 NME (64%), Class 1 (70%) and submissions for new indications (71%). Line extensions and generics received a positive decision in nearly all cases (>95%). For all submissions (n=824) for 'non-generic' products, the decision is preceded by evaluation of the therapeutic value as compared to the available alternatives: the outcome of this evaluation is either added or similar therapeutic value.

Whether 'evidence' is a significant factor affecting the reimbursement decision after the implementation of the above-mentioned EBM procedures, was indirectly investigated by analyzing the correlation between the decision made and the approval of the added value, for Class 1 submissions. The added therapeutic value claim was accepted in less than half (32 of 67) of all Class 1 submissions. The approval of the Class 1 label was highly predictive of a positive decision supporting the hypothesis that increasing the level of evidence – for added value at least one positive randomized active control RCT with superiority design is needed – facilitates a positive decision (Pearson Chi-Square = 12,26 p<0,001).

It appeared that the approval of added value was a strong factor affecting the reimbursement decision: the odds of a negative decision increased significantly when the added value claim was rejected (O.R. = 9,1 [2,3 – 35,6]). Indeed, 90% of all submissions labeled as Class 1 obtained reimbursement versus 50% of submissions where the evidence of therapeutic superiority was lacking. This shows that the reimbursement procedure facilitates the access of new innovative agents with good evidence of added therapeutic value.

It is important to specify that for Class 1 submissions a review of the therapeutic value is performed within 18 to 36 months in order to consider the accumulated experience post marketing and to assess whether the superiority primarily based on efficacy safety analysis also translates in a superiority in effectiveness.

Because in the Belgian procedure added therapeutic value is necessary to obtain a higher reimbursement basis, if the added value claim is rejected, some applicants may propose a lower reimbursement basis – similar to the existing alternatives – in order to get on the list of reimbursable medicines.

The median time to arrive at a decision by the Minister was 175 days: nearly all decisions (90%) were taken within 180 days but it took about an additional two months before the decision is really in application. The type of submission had no substantial impact on the time lines: median time to decision and application ranges respectively from 175 to 178 days and from 228 to 240 days. Before 2002, the mean time to a decision on reimbursement after registration was as high as 595 days [19]. The introduction of the new procedure was associated with a dramatic reduction of these time lines, which are now in compliance with the European Directive. The duration of the procedure of NMEs with a thorough analysis of the available evidence was not significantly longer than that of e.g. generics, indicating that the introduction of evidence based analysis does not prolong the procedure.

In summary, the analysis of the impact of evidence on the outcome of the reimbursement procedure – the decision to approve or refuse reimbursement for a medicinal product – for all files handled in the period 2002-2004, indicated that the demonstration of added therapeutic value is significantly associated with the decision. The impact of the factor 'added therapeutic value' on the outcome represents a clear incentive to develop new drugs addressing unmet or partially met needs. Moreover, introducing evidence-based reimbursement did not jeopardize the respect of strict EU time lines [20].

Orphan Medicinal Products

Following the successful Orphan Drug Act in the USA [21,22], the EU adopted its 'Regulation on Orphan Medicinal Products' in April 2000 to promote the development of drugs for patients suffering from rare diseases, the so-called orphan drugs[23,24]. In the EU, an Orphan Medicinal Product (OMP) designation can be obtained if the product is intended for the diagnosis, prevention or treatment of a life-threatening or chronically debilitating condition that affects fewer than five per 10,000 patients, or for which without incentives, it is unlikely that expected sales of the medicinal product would cover the investment in its development.

Another condition is that no satisfactory method for the diagnosis, prevention or treatment exists, or if such exists, that the new product will be of significant benefit for those affected by the condition[23]. An OMP designation does not automatically mean that the orphan drug will be authorized for marketing. Indeed, like other medicinal products, they have to comply with the criteria of quality, safety and efficacy.

An OMP designation may be obtained at any stage of development of the product and entitles the drug producer to several incentives, of which a market exclusivity of 10 years upon authorization that prevents any competitor of gaining market access unless superiority is demonstrated, is the most important. As in the USA, the European regulatory authorities also offer assistance in the design of clinical trials and they wave their normal licensing fees.

The (bio)pharmaceutical industry has responded to these incentives as evidenced by the increasing number of OMP designations and the increasing number of orphan drugs that receive market approval[24]. However, complete success can only be achieved if patients get access with timely reimbursement to therapies that are proven to increase their life expectancy and/or quality of life. Decisions regarding access and reimbursement are taken at national level[2] and, given the increasing financial pressure in health care and the high acquisition costs of orphan drugs, may vary between different EU countries[25].

We analyzed the reimbursement decisions of all orphan drugs submitted for reimbursement in Belgium from 2002 to 2007, with special emphasis on the quality of clinical evidence, as compared to innovative drugs for more common diseases.

A growing number of orphan drugs obtain market approval and are submitted for reimbursement

Rare or orphan diseases affect small numbers of patients but the number of diseases is very large – over 5,000 according to the World Health Organization (WHO). The total number of patients affected by orphan diseases was estimated to reach 30 million in Europe [24,26]. Hence, rare diseases are an important health issue and from a societal point of view, it is defendable that these patients should equally have access to safe and effective therapy, as do patients who suffer from more common diseases [27]. During the first six years of orphan drug legislation in Europe, the European Commission granted 442 OMP designations and approved marketing for 31 orphan drugs [24]. By August 2008 these numbers had increased to 519 and 44 respectively, indicating that the legislation and its incentives are increasingly successful in promoting R&D of drugs for rare diseases [24,28,29].

Between 1 January 2002 and 31 December 2007, 25 files of orphan drugs and 117 files of non-orphan innovative drugs (source: Administrative database of the National Insurance Agency) were submitted and evaluated by the CRM. During the first three years (2002-2004), orphan drug submissions represented only a small fraction of all submissions of new innovative compounds (six out of 71). During the second three-year period (2005-2007), however, orphan drug submissions increased to over a quarter of all submissions of new innovative medicines claiming added therapeutic value (19 out of 71).

Although most rare diseases are genetic in origin, only eight submissions involved hereditary metabolic diseases with only four drugs representing enzyme replacement therapy. Three orphan drugs were for pulmonary hypertension and 10 for hemato-oncological indications. Whereas the orphan drug legislation aims at promoting the development of drugs for rare diseases adequately targeted to patients with unmet medical needs, i.e. for which no therapy is yet available [23], only seven indications were not previously treated with another drug with at least partial efficacy.

Orphan drug designation predicts reimbursement

The CRM proposed reimbursement in 16 files, but did not reach a two-thirds majority for either a positive or negative proposal in five other files. The Minister of Social Affairs decided to reimburse all these orphan drugs. Three orphan drugs were submitted for indications for which a treatment was already available at a much lower cost and without any additional therapeutic benefit, and did not obtain reimbursement. A fourth orphan drug was also not accepted by the CRM because of the treatment cost (€350,000 per patient per year), and the absence of clinical evidence in patients younger than five years for whom the medical need is most important, but the Minister approved product reimbursement.

Therefore, 22 out of 25 (88%) submissions of orphan drugs resulted in reimbursement, in each case at the (high) price level requested by the company. The outcome of the reimbursement decision is clearly more often positive for orphan drugs than for the 117 other medicines claiming added therapeutic value, of which only 74 (63%) obtained reimbursement, often at a lower price level. It therefore appears that the orphan drug status is a strong predictor of reimbursement, and that orphan drugs gain market access easily in Belgium, despite the high cost and the fact that at the time of evaluation they do not have the same breadth and quality of clinical evidence as required for new medicines for more common diseases.

Poor quality of clinical evidence of orphan drugs

A quality analysis of the clinical evidence in all Belgian orphan reimbursement files from 2002 to 2007 confirmed and extended previous observations from a review of the EPARs of 18 orphan drugs that received marketing approval between 2000 and 2004[30]. The clinical data in many files leave many questions regarding long-term effectiveness versus safety and optimal dose unanswered. Randomized double blind active-controlled trials were present in one file and open controlled trials in two files only, although an active comparator was available in several cases. Only 12 files included randomized placebo-controlled trials. In one case, the comparison with historical control appears acceptable as this drug proved to be largely beneficial in terms of survival, but in many other cases the clinical data were less convincing.

Assessing the therapeutic value of orphan drugs may be more difficult than of other innovative medicines due to the rarity of patients, the disease's severity and heterogeneity and the scarcity of clinical experts[31]. However, in certain cases adequate placebo- or active-controlled trials were feasible, and the very small number of subjects included is not justifiable. Apart from some exceptions, the number of trial patients was generally inadequate. In 10 cases, less than 100 and in 16 cases less than 250 patients were included. Adequate dose finding studies were often lacking and, particularly for metabolic diseases, data generated in small children were extrapolated to adults without adjustment for disease severity and onset. In half of the files, the primary endpoints were surrogate endpoints with very little evidence of a beneficial effect on the clinical outcome. Effects on clinical endpoints were available in eight of the 10 files of orphan drugs for hemato-oncological indications, but only in four of the other 15. Moreover, in most cases the duration of drug exposure was far too short in relation to the natural history of the disease. Clearly, lower levels of clinical evidence for granting reimbursement and providing access to the therapy are more easily accepted than those required for other innovative medicines, where the decision more clearly reflects the amount and quality of the evidence [20]. For orphan drugs, a much higher degree of uncertainty regarding clinical effectiveness and safety appears acceptable.

Proof of added therapeutic value and cost-effectiveness required for reimbursement of innovative medicines for more common diseases but not for orphan drugs

As mentioned above, the introduction of EBM principles in the medicine reimbursement process in Belgium was a significant factor affecting the reimbursement decision [20]. Indeed, 90% of all new medicines for common diseases with clinical evidence of added therapeutic value, proven by adequate randomized active-controlled clinical trials, obtained reimbursement versus only 50% of the submissions where evidence of therapeutic superiority was lacking. In the latter cases, reimbursement was conditional on a price reduction to the price level of the standard alternatives available or the company did not apply for a price premium. In some cases the reimbursement decision was negative despite demonstrated added therapeutic value because of other factors such as the perception of excessive price, substantial budget impact or lack of cost-effectiveness.

A pharmaco-economic analysis is mandatory to obtain reimbursement at a price premium for non-orphan innovative medicinal products, but not for the approval of orphan drugs. It is obvious that if the same requirements and criteria would be used for orphan drugs, most would not obtain reimbursement because of the poor quality of the clinical evidence, the high cost per patient and the huge incremental cost per QALY[32]. The perceived societal value of orphan drugs is clearly a more determinant factor than the incremental cost-effectiveness in Belgium. Orphan drugs are apparently given more priority than other drugs for equally severe but more common conditions.

Access to orphan drugs

A balance must be found enabling rapid access to orphan drugs while guaranteeing their quality, efficacy and safety. Less rigid criteria are used for the clinical evaluation of the therapeutic value of certain orphan drugs[31,32]. Enrolling sufficient trial patients can be difficult in rare diseases. Nevertheless, our analysis indicates that even taking into account the difficulties associated with smaller patient populations and disease heterogeneity, better designed trials and longer duration of exposure to more patients could have been achieved in many cases. For diseases considered orphan, based on prevalence criteria because of short survival despite a relatively common incidence, classical types of randomized trials with clinical endpoints are feasible[33].

Regulators bear the responsibility for ensuring that patients are not denied vital therapy, but also for protecting public health by ensuring that medicines made available are effective, safe and of adequate quality. Therefore, there is a need for more quality clinical data to help regulators and payers in assessing the risks and benefits of orphan drugs. In many cases these data are missing, although at least in some cases more robust evidence could have been provided. Some argue that the budgetary impact is limited given the small number of patients[34]. However, although very rare conditions individually equate to small patient numbers, there are cumulatively a large number of patients with these diseases[24,26]. The increasing fraction of the available health care resources devoted to orphan diseases implies that in a resource limited environment like Belgium, fewer resources will be available for more common diseases and for larger groups of patients. It is defendable that payers require guarantees that the resources they allocate to orphan diseases are well spent. One can question whether the health care system should pay more for a treatment than the value for health care

[51]

it produces, even if it is an orphan drug. Some authors propose that cost-effectiveness should not be the only determinant of access to orphan drugs and that a QALY weighting for disease rarity could be used [32,34,35], but others have argued that rarity alone is not a good argument for a special status and incompatible with equity principles [36,37]. It indeed remains to be shown whether society really places a higher value on health gains achieved in patients with rare diseases compared with more common disorders.

Orphan drugs made available to patients despite uncertainty on their effectiveness and safety, should be part of a prospectively designed structured program of post-marketing surveillance, and the treatment should be stopped if not producing benefit. Conditional pricing and reimbursement linked to the creation of registries of all patients receiving the treatment, and well coordinated within and between the Member States, could be a way to reconcile the need for access to treatment for patients with rare diseases with the necessity to assure that these treatments are effective and safe, and that the resources are well allocated.

Conclusion

In order to be compliant with the European Transparency Directive, Belgium adapted its reimbursement process by integrating EBM principles in the evaluation process. This had a significant impact on the outcome of reimbursement decisions as demonstrated by the fact that new medicines for common diseases with good clinical evidence of added therapeutic value, proven by adequate randomized active-controlled clinical trials, obtain reimbursement more often than submissions where evidence of therapeutic superiority was lacking.

Orphan drugs obtain reimbursement in the majority of cases despite a lower quality of clinical evidence, a higher level of uncertainty on the clinical effectiveness, safety and incremental cost-effectiveness, and a higher budgetary impact than new innovative drugs for more common diseases. This suggests that health authorities value the benefits of orphan treatment more highly compared to benefits of treatments of equally severe more common diseases. Further research should be done to examine whether a societal preference for such an 'orphan' premium really exists.

BIO

Alain Dupont, MD, PhD, is full professor and head of the Department of Pharmacology at the Vrije Universiteit Brussel, and of the Department of Pharmacology and Pharmacotherapy at the University Hospital UZ Brussel, where he also runs a hypertension clinic. He was Vice-Dean Research and is currently Dean of the Faculty of Medicine and Pharmacy. He is founding president and member of the Belgian Commission for Reimbursement of Medicines. His research activities, focused on hypertension and cardiovascular pharmacology, resulted in three prizes and over 120 publications in peer reviewed journals and books. He chairs the Accompanying Committee of the CROSSTALKS project 'The Future of Medication'.

REFERENCES

1. Commission of the European Communities. Transparency Dir.2001/83/EC of the European Parliament and of the Council. Off J Eur Commun. 2001; L311:67-128.
2. Commission of the European Communities. Transparency Dir. 89/105/EEC of the European Council. Off J Eur Commun. 1989; L40:65-71.
3. RIZIV, Rijksinstituut voor Ziekte en Invaliditeitsverzekering, Geneesmiddelen en andere verstrekkingen. http://inami.fgov.be/drug/nl/drugs/reglementation/ar-20011221/pdf/arkb20011221.pdf
4. Haute Autorité de Santé. www.has-sante.fr/portail/upload/docs/application/pdf/ri_ct_2005_v.04-10-06.pdf
5. College voor Zorgverzekering, Procedure aanvraag vergoeding geneesmiddelen: beoordelingscriteria. www.cvz.nl/resources/CFH_beoordelingscriteria_tcm28-15811.pdf
6. National Institute for Clinical Excellence. www.nice.org.uk/page.aspx?o=201971
7. International Conference on Harmonisation. www.ich.org/cache/compo/276-254-1.html
8. European Medicines Evaluation Agency, Guidelines. www.emea.europa.eu/htms/human/humanguidelines/efficacy.htm
9. Rothwell PM. Treating Individuals 1.External validity of randomised controlled trials: "To whom do the results of this trial apply?". Lancet. 2005; 365: 82-93.
10. Gøtzsche PC. *Believability of relative risks and odds ratios in abstracts: cross sectional study.* BMJ. 2006; 333: 231-4.
11. MacMahon S, Colins R. *Reliable assessment of the effects of treatment on mortality and major morbidity, II: observational studies.* Lancet. 2001; 357: 455-62.
12. Horn SD. *Performance Measures and Clinical Outcome.* JAMA. 2006; 296(22): 2731-2.
13. International Society for pharmacoepidemiology, Resources, ISPE Policies and Presentations. www.pharmacoepi.org/resources/policies.cfm
14. College voor Zorgverzekering, Procedure aanvraag vergoeding geneesmiddelen. www.cvz.nl/resources/CFH_procedure_aanvraag_vergoeding_geneesmiddelen_tcm28-15809.pdf
15. Joppi R, Bertele' V, Garattini S. *Disappointing Biotech.* BMJ 2005;331:895-7.
16. Haute Autorité de Santé, *Rapport d'activité HAS 2005*, p.18. www.has-sante.fr/portail/display.jsp?id=c_443237
17. Charles Rivers Associates, *Innovation in the Pharmaceutical Sector*, Nov.8th 2004
18. Food and Drug Administration, *Challenge and Opportunity on the Critical 19. Path to New Medical Products.* www.fda.gov/oc/initiatives/criticalpath/whitepaper.pdf
19. Pharma.be, Factua Newsletter 2001;118:1-1. www.pharma.be/data/File/factua%20newsletter/factua%20newsletter%20archives/nl/fa118_nl.pdf

20 Van Wilder PB, Dupont AG. *Introducing evidence-based medicine in reimbursement procedures: does it affect outcome?*, Value Health 2008; 11: 784-7.
21 Federal Drug Administration. The Orphan Drug Act. Public Law 1983; 97-141. www.fda.gov/orphan/oda.htm.
22 Haffner ME, Whitley J, Moses M. *Two decades of orphan product development.* Nat Rev Drug Discov 2002; 1:821-5.
23 European Union. Regulation (EC), No. 141/2000 of the European Parliament and of the Council on Orphan Medicinal Products of 16 December 1999 and Commission regulation (EC) No 847/2000 of 27 April 2000. Off J Eur Comm. 2000; 43: L103/5-8.
24 Haffner ME, Torrent-Farnell J, Maher PD. *Does orphan drug legislation really answer patients' needs?*, Lancet 2008; 371:2041-44.
25 Alcimed. *Study on orphan drugs. Overview of the conditions for marketing orphan drugs in Europe.* Paris: Alcimed: 2005.
26 Wästfelt M, Fadeel B, Henter JI. *A journey of hope: lessons learned from studies on rare diseases and orphan drugs.* J Intern Med 2006; 260: 1-10.
27 Schieppati A, Henter JI, Daina E, Aperia A. *Why rare diseases are an important medical and social issue.* Lancet 2008, 371.2039-41.
28 Miles KA, Packer C, Stevens A. *Quantifying emerging drugs for very rare conditions.* Q J Med 2007; 100: 291-5.
29 European Agency for the Evaluation of medicinal products. Register of designated Orphan Medical Products. http://ec.europa.eu/enterprise/pharmaceuticals/register/orphreg.htm
30 Joppi R, Bertele V, Garattini S. *Orphan drug development is progressing too slowly.* Br J Clin Pharmacol 2006; 61: 355-60.
31 Buckley BM. *Clinical trials of orphan medicines.* Lancet 2008; 371: 2051-55.
32 Drummond MF, Wilson DA, Kanavos P, Ubel P, Rovira J. *Assessing the economic challenges posed by orphan drugs.* Int J Technol Assess Health Care. 2007; 23: 36-42.
33 Dear JW, Lilitkarntakul P, Webb DJ. *Are rare diseases still orphans or happily adopted? The challenges of developing and using orphan medicinal products.* Br J Clin Pharmacol 2006; 62: 264-71.
34 Hughes DA, Tunnage B, Yeo ST. *Drugs for exceptionally rare diseases: do they deserve special status for funding?*, Q J Med 2005; 98: 829-36.
35 Hughes DA. *Rationing of drugs for rare diseases.* Pharmacoecon 2006; 24: 315-6.
36 Mc Cabe C, Claxton K, Tsuchiya A. *Orphan drugs and the NHS: Should we value rarity?*, BMJ 2005; 31:1016-9.
37 Mc Cabe C, Tsuchiya A, Claxton K, Raftery J. *Orphan drugs revisited.* Q J Med 2006; 99: 341-5.

THE INCREASING NEED FOR INDEPENDENT RESEARCH

SILVIO GARATTINI

Not everybody realizes that most regulatory policy works in favor of pharmaceutical companies. The EMEA, the agency that approves new drugs for all the Member States of the European Union (EU) depends – unlike in most single countries – on the Directorate of Industry, suggesting that drugs are considered more as consumer goods than tools to cure diseases.

Reinforcing this impression, European legislation requires that all new medicinal products must be approved on the basis of quality, efficacy and safety, three important characteristics, but omits to require any added value. There is therefore a reasonable risk that new drugs are less effective – and often more costly – than those already on the market. This situation makes it difficult for the Member States to decide which drugs should be reimbursed by their national health services.

The lack of comparative data

The lack of comparative data leaves room for all sorts of pressure from industry, scientific societies and patients' organizations to introduce new drugs 'automatically' into national formularies. Even more surprising is the fact that the entire documentation presented for approval is conceived, produced and assembled by the pharmaceutical company that will market the product. If on top of that we add that the whole dossier for drug approval is protected by confidentiality, it soon becomes clear that it is very difficult to exert any control[1].

Several analyses indicate that biases tend to pervade the planning, conducting and results of the controlled trials necessary to assess efficacy and toxicity of drugs. Although EMEA guidelines require at least two pivotal comparative phase III trials, very few products fulfill this requirement[2]. There is quite a discrepancy between what is written in the protocol and what is reported in publications. More than 50% of the primary outcomes indicating benefit or harm are ignored or changed in the final report[3]. Most studies show long lists of conflicts of interest[4].

Papers of industrial interest are frequently written by ghost writers and analyzed by ghost statisticians[5]. The design of controlled trials involves placebo even when active drugs are available for the same indications[6]. When comparators are employed, they are not always the best drug or at the optimal dose and treatment schedule. Examples are the selection of NSAIDs as comparators for coxibs[7]. Mycophenolate, presented as a considerable improvement for the treatment of organ rejection[8], was recently found to be comparable to the older drug azathioprine in terms of efficacy and safety, though not price[9]. Many new drugs are studied with equivalence or non-inferiority designs rather than superiority, a choice that does not establish the drug's real role in the therapeutic armamentarium[10]. It is also ethically questionable to use patients in this kind of design whose sole purpose is to obtain a slice of the market[6]. In addition, what is presented as equivalent is frequently an excuse not to look for a difference[11, 12].

Juggling with the side effects

Another potential bias in the evaluation of a drug is the focus on certain side effects, glossing over others: a typical case is the second-generation anti-psychotics. These drugs may be better tolerated as far as the extra pyramidal side effects are concerned, but this advantage is unfortunately compensated by body weight gain, cholesterolemia and glycemia, which are important risk factors for diabetes and cardiovascular diseases[13]. Similarly, for the anti-inflammatory coxibs the emphasis was laid on reduced gastro toxicity, at the expense of myocardial infarction and heart failure[14]. Ironically, because of physicians' confidence in coxibs, English hospitals have seen an increase in cases with gastrointestinal bleeding[15]. It is also curious that the same drug seems more efficacious when it is new than when it is used as a comparator, as shown by the case of doxorubicin[16] and fluoxetine[17].

Publication selectivity

The perception of drug efficacy is additionally confounded by publication selectivity because positive trials have about three times more chance of being published than negative ones [18,19]. Furthermore studies conducted or supported by industry are more likely to favor the new drug than trials by non-profit groups [20]. The activity of donepezil in the treatment of Alzheimer disease was questioned by the independent trial AD2000 [21]. The Allhat study on antihypertensive agents showed that α-adrenergic inhibitors have an unfavorable benefit-risk ratio compared to other antihypertensive drugs, while the old diuretics have been shown to be substantially better than more recent drugs [22]. Special problems arise with pharmacovigilance because the regulatory agencies do not have the means to monitor drug toxicity, as was shown by the fact that cerivastatin and rofecoxib were withdrawn by the companies rather than by the EMEA or FDA.

All the considerations and examples given above call for new initiatives that will avoid drug approval relying solely on data produced by the pharmaceutical industry. A recent BMJ editorial suggested that at least one pivotal phase III trial should be carried out by independent clinicians [23].

Independent research

Whatever the specific initiatives, there is clearly a need to increase support for independent research. Clinical trials must be conducted, independently of the source of financial support, in such a way that authors have complete control of data analysis and publications. In this respect it may be of interest to report the experience of the newly established Italian Agency for Drugs (AIFA), whose role is to draft the list of drugs reimbursed by the National Health Service (NHS). A recent Italian law requires all pharmaceutical companies operating in Italy to pay a fee every year of 5% of their promotional expenses, except salaries [24]. This money must be utilized to support independent clinical research in three areas: efficacy of orphan drugs, head-to-head comparison of drugs with the same indications, studies of outcome and pharmacovigilance. Studies are aimed at collecting data usually not provided by pharmaceutical companies, but that may be important for decisions related to the admission or maintenance of drugs for NHS reimbursement. To implement this law a ten-member committee has been set up, half nominated by AIFA and half by the Regions. This Committee, after suitable consultations with the technical committee of the AIFA and other scientific institutions, launched a call for proposals in July 2005 indicating topics of interest. The selection was made in two steps. First, the Committee checked that the projects presented were consistent with the aim of the law. Out of 404 letters of intent, 98 were selected to present a full proposal. These proposals were made available to three international study groups that made a written evaluation, then met in Rome for three days and drew up a list of priorities, using a scoring system. In March 2006, 54 projects were approved for a total cost of about €35 million. For Italy this procedure offered an unprecedented independent evaluation for public financing of research. In July 2006 all the contracts were signed and work could start. Examples of approved projects include evaluation of orphan drugs for various rare diseases, comparison of ACE inhibitors and sartans, evaluation of three aromatase inhibitors for breast cancer, modalities of use of monoclonal antibodies for cancer treatment, etc.

In July 2006 a second call for proposals was approved by AIFA and launched after consultation with interested parties. Other topics have been considered, still in the three areas mentioned above. This second call attracted encouraging participation by Italian clinicians. Out of 454 letters of intent 104 projects were selected and 51 projects were approved by the study groups at the end of March 2007. The approved projects and topics considered by the calls for proposals are listed on the website[25]. In July 2007 a third call was launched following the same procedures resulting in the approval of 49 projects.

Structural support for independent clinical trials

To our knowledge this is the first time in Europe that a regulatory agency is supporting a clinical research program entirely devoted to the independent evaluation of drugs already on the market, to satisfy public health needs. This program should be followed by other European countries too, and particularly the EU, which has never seriously considered supporting independent clinical trials. The availability of independent research may be one way to improve clinical studies in Italy. Finally, independent research may counterbalance the weight of industry's clinical trials, resulting in better appraisal of drug efficacy and toxicity.

REFERENCES

1. Garattini S, Bertele V. *Adjusting Europe's drug regulation to public health needs.* Lancet 2001; 358:64-7.
2. Apolone G, Joppi R, Bertele V, Garattini S. *Ten years of marketing approvals of anticancer drugs in Europe: regulatory policy and guidance documents need to find a balance between different pressures.* Br J Cancer 2005; 93:504-9.
3. Chan AW, Hrobjartsson A, Haahr MT, Gotzsche PC, Altman DG. *Empirical evidence for selective reporting of outcomes in randomized trials: comparison of protocols to published articles.* JAMA 2004; 291:2457-65.
4. Remuzzi G, Schieppati A, Boissel JP, Garattini S, Horton R. *Independent clinical research in Europe.* Lancet 2004; 364:1723-6.
5. Hargreaves S. *Ghost authorship of industry funded drug trials is common, say researchers.* BMJ 2007; 334:223.
6. Garattini S, Bertele V, Li Bassi L. *How can research ethics committees protect patients better?* BMJ 2003; 326:1199-201.
7. Psaty BM, Weiss NS. *NSAID trials and the choice of comparators – questions of public health importance.* N Engl J Med 2007; 356:328-30.
8. The Tricontinental Mycophenolate Mofetil Renal Transplantation Study Group. *A blinded, randomized clinical trial of mycophenolate mofetil for the prevention of acute rejection in cadaveric renal transplantation.* Transplantation 1996; 61:1029-37.
9. Remuzzi G, Lesti M, Gotti E, Ganeva M, Dimitrov BD, Ene-Iordache B, et al. *Mycophenolate mofetil versus azathioprine for prevention of acute rejection in renal transplantation (MYSS): a randomised trial.* Lancet 2004; 364:503-12.
10. Splawinski J, Kuzniar J. *Clinical trials: active control vs placebo – what is ethical?* Sci Eng Ethics 2004; 10:73-9.
11. Barbui C, Violante A, Garattini S. *Does placebo help establish equivalence in trials of new antidepressants?* Eur Psychiatry 2000; 15:268-73.
12. Bertele V, Torri V, Garattini S. *Inconclusive messages from equivalence trials in thrombolysis.* Heart 1999; 81: 675-6.
13. Lieberman JA, Stroup TS, McEvoy JP, Swartz MS, Rosenheck RA, Perkins DO, et al. *Effectiveness of antipsychotic drugs in patients with chronic schizophrenia.* N Engl J Med 2005; 353:1209-23.
14. Juni P, Nartey L, Reichenbach S, Sterchi R, Dieppe PA, Egger M. *Risk of cardiovascular events and rofecoxib: cumulative meta-analysis.* Lancet 2004; 364:2021-9.

15. Mamdani M, Juurlink DN, Lee DS, Rochon PA, Kopp A, Naglie G, et al. *Cyclo-oxygenase-2 inhibitors versus non-selective non-steroidal anti-inflammatory drugs and congestive heart failure outcomes in elderly patients: a population-based cohort study.* Lancet 2004; 363:1751-6.
16. Fossati R, Confalonieri C, Apolone G, Cavuto S, Garattini S. *Does a drug do better when it is new?* Ann Oncol 2002; 13:470-3.
17. Barbui C, Hotopf M, Garattini S. *Fluoxetine dose and outcome in antidepressant drug trials.* Eur J Clin Pharmacol 2002; 58:379-86.
18. Stern JM, Simes RJ. *Publication bias: evidence of delayed publication in a cohort study of clinical research projects.* BMJ 1997; 315:640-5.
19. Melander H, Ahlqvist-Rastad J, Meijer G, Beermann B. *Evidence b(i)ased medicine – selective reporting from studies sponsored by pharmaceutical industry: review of studies in new drug applications.* BMJ 2003; 326:1171-3.
20. Cho MK, Bero LA. *The quality of drug studies published in symposium proceedings.* Ann Intern Med 1996; 124:485-9.
21. Courtney C, Farrell D, Gray R, Hills R, Lynch L, Sellwood E, et al. *Long-term donepezil treatment in 565 patients with Alzheimer's disease (AD2000): randomised double-blind trial.* Lancet 2004; 363:2105-15.
22. Messerli FH, Grossman E. *Doxazosin arm of the ALLHAT study discontinued: how equal are antihypertensive drugs? Antihypertensive and Lipid Lowering Treatment to Prevent Heart Attack Trial.* Curr Hypertens Rep 2000; 2:241-2.
23. Godlee F. *Can we tame the monster?* BMJ 2006; 333.
24. Gazzetta Ufficiale della Repubblica Italiana. Capo IV. *Accordo Stato-Regioni in materia sanitaria. Art. 48: Tetto di spesa per l'assistenza farmaceutica.* 2003; Supplemento Ordinario n.274 (Parte prima, martedi' 25 novembre): 95-98.
25. Agenzia Italiana del farmaco-AIFA. *Bandi per la ricerca indipendente sui farmaci 2006.* www.agenziafarmaco.it/aifa/servlet/section.ktml?target=&area_tematica=INFO_SPER_RIC&cache_session=true&section_code=AIFA_BANDI_RICERCA_IND2006.

BIO

Silvio Garattini MD, PhD, is founder in 1963 and director of the Mario Negri Institute for Pharmacological Research. Professor Garattini is also the founder of the European Organization for Research and Treatment of Cancer (EORTC), chairman of the UICC Committee on Antitumoral Chemotherapy and the European Society of Biochemical Pharmacology. He served as a member of the Italian National Research Council (CNR) Committee on Biology and Medicine, the National Health Council, the Committee for Italian Research Policy, set up by the Presidency of the Council of Ministers, the Commissione Unica del Farmaco (CUF), and is a consultant to the World Health Organization. He is currently a member of the Committees for Proprietary Medicinal Products (CPMP) of the European Agency for the Evaluation of Medicinal Products (EMEA), member of the CEPR (Committee of Experts of Research Policy) at the Ministry for the University and Scientific and Technological Research, President of the V Section of the Consiglio Superiore della Sanità, Rome, and member of the Board of Istituto Superiore di Sanità, Rome. Silvio Garattini is a Fellow of the New York Academy of Sciences, the American Association for the Advancement of Science, Honorary Fellow of the Royal College of Physicians (Pharmaceutical Medicine), London, and a member and President of numerous other Italian and international scientific societies. He has received many awards for his work, including the French Legion d'Honneur and the Commendatore della Repubblica Italiana, and holds honorary degrees from the Universities of Bialystok in Poland, and Barcelona in Spain. Garattini's publications both in Italian and English appear in international scientific journals, and his texts on pharmacology run into the hundreds.

THE COST AND ADDED VALUE OF PERSONALIZED MEDICINE

JACQUES DE GRÈVE

In the last decade, a number of novel treatments have become available that sometimes have a spectacular, but often small impact on the outcome of the cancer patients on which they are used. These treatments are called 'targeted therapies', because they interact with specific proteins in the cancer cell. They build on fundamental insights into the biology of the cancer cell and in particular the 'cancer genes' that have been acquired in the last thirty years.

In brief

In the laboratory, these drugs are very successful at killing cancer cells or inhibiting their growth if their target is 'constitutionally activated' by a genomic alteration (for example a mutation) of the gene that encodes the target protein. In addition we also know specific genomic alterations that generate high resistance to some of these drugs. These drugs thus have specific mechanism(s) of action that could provide a sound scientific basis for their use in patients who have the same genomic alteration as the ones found to correlate with activity in the laboratory.

Despite this, these drugs have been uniformly developed with methods that ignore this selective efficacy and the clinical studies have been performed in unselected patient populations. The results of these studies then serve as a basis for registration with the European and American authorities and this subsequently forms the basis for reimbursement of these medicines by the authorities or insurance companies. As a consequence these drugs are currently not adequately used.

It would be very important to target these drugs at the patients who benefit most from these treatments for two major reasons:

1. Patients who receive the drug and do not benefit still suffer toxicities and possibly even deleterious effects on survival.
2. The most pressing reason is that these drugs are very expensive and that using them in non-benefitting patients leads to a huge waste of resources.

A third, academic, reason is that the patients who do not receive the drug with scientific justification are also, just because they are receiving these futile treatments, not flowing into other clinical trials that could possibly benefit them or at least advance clinical science.

The issues surrounding these molecular treatments are dual: the high price and the patient selection. We will try to demonstrate this with the help of an exemplary case study: targeted therapies in lung cancer.

Introduction and context of these new treatments

Cancer treatments are local (surgery and radiotherapy) or systemic (hormonal treatment in breast and prostate cancer, chemotherapy and/or the novel molecular targeted therapies in all cancers).

The novel molecular therapies target specific proteins that play a role in cancer biology. These proteins are either the products of oncogenes (activated genes) or tumor suppressor genes (inactivated genes). The therapeutic reactivation of tumor suppressor genes has been explored and there is some clinical *proof of principle*, but today there are still important technological limitations that need to be solved.

The most impressive successes with regard to cancer treatment have been achieved by targeting the proteins that are encoded by activated oncogenes in the cancer cell and by inhibiting this inappropriate function in various ways.

There are two categories of targets (see Figure 1):

1. Products of cancer genes in the cancer cell.
2. Proteins in surrounding tissues (stroma) and, most importantly, proteins involved in angiogenesis that have been indirectly activated by signals coming from the cancer cell. Indeed, the behavior of the cancer in a patient depends on the interaction of these cancer cells with their surrounding stroma, including the capacity to stimulate the formation of new blood vessels.

Economic aspects

These novel drugs uniformly have a high price and many of them situate in a bracket of €25,000-50,000/year per treatment. A high price could be perfectly justified for innovative first-in-class drugs as the clinical development through large comparative phase III studies is very expensive and has become even more so due to heavy regulation (and overregulation). Some of these phase III studies are also burdened by the gathering of details that are not really important for the progress (improving survival) in cancer treatment.

The cumulative cost of phase III studies is in the billion euro range. The high price is generated by a number of visible and invisible factors including the relative short protection period for these innovative drugs after the end of the studies, which take several years to complete. The expense also includes the basic science that has lead to the discovery of the target and the invention of the drug. It should be mentioned that most of the initial research on which these treatments are based is done in the academic world and this effort does not always receive a commensurate financial reward. While a high price could be justified for these innovative drugs, it is less justified for next-in-line drugs that target the same protein and are similar to the first-in-line drugs.

Patient selection

The second issue is the patient selection, which has become not only a financial but also a medical imperative. It can be stated that what we do in the clinic right now is much less than what science would allow doing as far as patient selection is concerned. In fact, the current selection criteria not only fail to restrict sufficiently the patient population treated, but also sometimes exclude patients who could derive great benefit and would be recognized if genomic methods were used.

The case study: Lung cancer

Lung cancer is a primary cause of death in the male age group of 50 to 70 years and in Europe was responsible for 334,800 cancer deaths in 2006 or 19,7 % of total cancer deaths. Eighty five percent of these cases are non-small cell lung cancers (NSCLC). The median survival rate of these cancers is eight months in an advanced stage and this includes modern chemotherapy and radiotherapy. So we definitely need improved therapies. Until 2005 there were no such treatments available.

Two important targets are being used in the treatment of lung cancer: pathways of angiogenesis and pathways within the cancer cell (Figure 1).

Figure 1

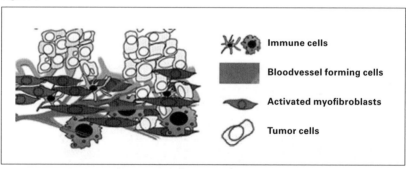

Tumors are composed, not of only tumor cells, but also normal cells (called stroma) that communicate with the tumor cells and enhance their malignant behavior. Tumors also secrete growth factors that stimulate new blood vessel formation (angiogenesis) needed for tumor growth. Modern therapeutic agents can target proteins in the tumor cells or in the stroma or both.

Targeting angiogenesis

Cancer tumor cells secrete growth factors that stimulate the formation of new blood vessels. One of these therapeutic agents is bevacizumab (Avastin®), which is directed against one of these important growth factors and neutralizes it, thereby blocking and reversing blood vessel formation. When this antibody is used together with chemotherapy in advanced non-small cell lung cancer, the response to the chemotherapy is more than doubled, 35% compared to 15% using chemotherapy alone. The median survival rate is increased by two months. It is suggested that by normalizing the blood vessels within the tumor, the chemotherapy is better able to penetrate the tumor.

[69]

The cost however is tremendous, for six cycles (about five months of treatment) the additional cost of adding bevacizumab to chemotherapy is €12,500. Most of these patients get five additional treatments after chemotherapy at a cost of €10,500 or a total of €22,000 per patient. When we compute this to the survival advantage we come to a rough estimated cost of €130,000 per life year gained. Calculated in the Belgian setting this would have a budgetary impact of €72,000,000.

Figure 2

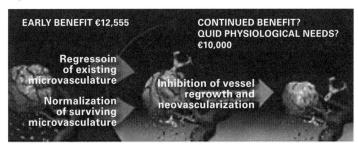

Inhibition of angiogenesis improves significantly the efficacy of chemotherapy when given simultaneously. In the clinical studies and in practice angiogenesis blocking drugs are continued beyond the administration of chemotherapy. The validity of this (costly) strategy has not been addressed in the clinical studies. It might well be that limiting the duration of such treatment is the better strategy from both a medical and an economical point of view.

How can we limit the cost?

Biomarkers (parameters in the patient or the tumor) that could predict the usefulness of the treatment are lacking. This is not surprising since the angiogenesis inhibitors work in a very complex system: interaction between the cancer cell and the host. Probably several variables will determine whether this treatment is successful or not. In clinical studies it has been possible to negatively select patients at excessive risk and these patients are excluded from such treatment, but about half of the patients remain eligible.

Reducing the duration of treatment could result in possible savings. If the treatment is not continued beyond chemotherapy (which would be reasonable in view of the mechanism of synergy) then that would represent a cost reduction of 44% or €73,000 per life year gained and a budget impact of (only) €40,000,000. Unfortunately, the optimal duration of the treatment has not been studied in any of the phase III trials that led to the registration of this drug for use in several cancers. Longer continuation of the drug has been justified by the observation that in mice once the drug is stopped, blood vessels start to re-grow. However, humans are not mice and the biology between the two is significantly different.

So for example, it could be possible that continuation of the drug *might* be beneficial for its antitumor effects but detrimental for human physiology. Indeed, we do not have sufficient insight into what angiogenesis inhibition does to the patient outside of the tumor. So, what should have been done are trials in which the duration of treatment was also studied.

Market size considerations, however, stimulate companies to perform large-scale studies that not only fail to look for patient enrichment but also want to give the drug to as many patients as possible and for as long a time as possible. This is because the drug development scheme is almost entirely proposed and controlled by the pharmaceutical industry that generates these valuable drugs. And there is no alternative because academics and other organizations do not have the motivational driving force to complete such important trials in an acceptable time frame, but the design of the trials and the choice of biomarkers and endpoints should be much more influenced by academics and health authorities.

Targeting pathways in the cancer cell

The second example is when the target is in the cancer cell. An exemplary system is the Epidermal Growth Factor Receptor (EGFR). This receptor is a member of a four-receptor family of phylogenetic related receptors.

The EGFR and its ligand are expressed in lung cancer, sometimes at high levels. Many are misled by this high expression level and believe that this automatically means that such a protein is important for therapeutic targeting. However, it turns out that the normal progenitor cells from which the cancer has arisen also have the high levels of expression of the receptor and the normal lung tissues also produce large amounts of ligands for this receptor. So the mere presence of this protein and ligand system in the tumor does not mean that it contributed to the process of malignant transformation, but could very well be a physiological property of these cells.

The EGFR is a cell membrane receptor that after binding with a growth factor starts a signal transduction pathway that will ultimately regulate growth and survival of the cell. Experimentally we know that only cancers in which this pathway is activated, for example by mutation or over expression due to increased gene copy number, are dependent on it and can be exquisitely sensitive to inhibition of this target.

Small molecule inhibitors and antibodies have been developed against this receptor. The small molecules penetrate the cell and block the enzymatic activity of the receptor and thus block the signal transduction pathway. When such a drug, Erlotinib (Tarceva®), was given to patients after the failure of chemotherapy and compared to placebo, only a two-month increase in median survival in unselected NSCLC patient populations was found. However, some of these patients had dramatic long-term remissions.

Retrospective analyses of these broad phase III studies revealed that there is a clinical phenotype that predicts high activity of the drugs: adenocarcinoma in nonsmokers, a phenotype that occurs more often in Asians and females. In these patients, an impressive response rate of 60% can be obtained. Also biological biomarkers have been found: EGFR gene copy number and mutations that make the receptor abnormally active but also much more sensitive to inhibition by the inhibitor. Increased copy number *predicts* a response in half of the patients and a progression free survival of 6,4 months while a mutation *predicts* a response rate of 82% and progression free survival of 13 months. This is really impressive compared to pre-existing therapies.

Figure 3

	Mutant 5%	FISH 15%	WT 80%
Response	80%	50%	< 5%
T duration	14	7	2
Cost	122,000	184,000	280,000
%	21%	31%	48%
S gained	1,5 years	10 months	< 1 month
Cost/LYG	€16,250	€14,720	> €42,000

Cost of erlotinib treatment in three different genomically defined patient groups in relationship to the differential cost effectiveness. The figures in the table apply to a cohort of 100 patients. The majority of patients treated derive a minimal or no benefit at all and account for half of the total drug cost.

Taking into account all these data and the cost of Erlotinib, we can estimate that wild-type patients (75% of the patients), who have less than a 2% response rate and are only treated for an average of two months, account for almost half of the budget spent on Erlotinib. And this for an average survival gain of less than one month. The cost per life year gained is more than €40,000, three times as much as the cost of the same drug in patients with a sensitive cancer. In addition, several trials suggest that patients with a wild-type receptor not only do not benefit from this treatment but also might actually have a lower survival rate than when treated with a placebo.

In this case, it is evident that the drug should be used only in patients in whom the cancer cells have an activated target (20%-25% of the patients) and not in the patients in whom this activation can not be shown, not only for cost-effectiveness, but also for medical reasons. Such reasoning is in principle applicable to any other targeted therapy in any cancer.

The drug development strategy should be changed to accommodate these new drugs

The consequence of only relying on the outcomes of unselective large phase III studies for registration and reimbursement of new targeted agents, the system used to develop chemotherapy and other drugs, leads to a situation in which drugs that could work very well in very small subsets of patients who are hypersensitive to them reach clinical practice for several years or not at all. On the other hand, these expensive drugs are also used in patients who have no biomarker for predicting benefit, which leads to huge waste and is medically deleterious.

One could argue that for some unforeseen and as yet not scientifically revealed mechanism, some of these other patients in whom no predictive biomarkers are known could perhaps respond to such treatments. This is even more likely with so called *dirty* small molecules that target several proteins. There is thus a *possible* grey zone. The solution is to organize phase III studies in patient groups who do not have a predictive biomarker. The advantage of this approach will be that when such an unpredicted response would occur, these patients could then be investigated thoroughly for the genomic alteration to discover why these responses occur and thus foster scientific progress.

Proposal

We propose that the development and clinical application of these drugs should be based on pre-clinical science and that it should be sufficient to show high efficacy of such drugs in phase II studies in biomarker selected patients. In grey zones, phase III studies should/could be performed. Such a strategy could lead to faster (in years) access of patients to such novel treatments. To achieve this, the relevant authorities should create a context that makes this possible and academics and in particular pre-clinical scientists, should have greater influence on the design of studies. In order to engage the industry along these lines, it would be proper to consider a longer protection of medication specifically for application in patient groups in which they have a high value for money. The grey zones should perhaps be investigated (if judged worthwhile) by the industry/academics in shared financial responsibility with the insurance bodies.

Such an approach, selecting patients on biomarker prior to treatment, is practically feasible, as we have shown in a Belgian multicenter study (17 centers) in which Erlotinib was prospectively administered after prior identification of lung cancer patients for genomic sensitivity (mutations in the EGFR genes). The genomic analysis can be done in less than 10 to 14 days, which is fully compatible with clinical decision-making. Such a context has also allowed us to genomically analyze patients and reveal mutations in other genes that then allow treating these patients with other investigational agents that are predicted to be highly active on these alternative targets.

Not only public finances and patients would gain by a biomarker-based system, but also the industry. Indeed, currently these sensitive patients are retrospectively identified years after performing the phase III studies. In the meantime the price has been calculated based on a marked size that includes 100% of the patients, creating a huge problem when a *posteriori* science indicates that only 20% of the patients should receive the treatment. Identifying these 20% from the start would allow more appropriate market size calculation and hence a more adequate price setting, also from a corporate point of view.

Perspectives

Today we only know druggable targets in a small minority of all cancers. However, current technology is rapidly evolving that will allow us to fully analyze the genome of the cancer of any patient and to look for as yet unknown activated cancer genes that should each undergo proper functional scientific validation as a therapeutic target.

The model we are ultimately moving towards is that once the repertoire of activated cancer genes is identified in a specific patient, that this can then be matched to the repertoire of available drugs, in order to choose the most appropriate drug or drug combination for that patient. One important condition to be able to implement such a strategy is that the infrastructure to work genomically is progressively made available in all major cancer centers. In particular the introduction of this technology in molecular pathology labs should be an absolutely priority.

BIO

Jacques De Grève (MD, PhD) is a medical oncologist and head of the Medical Oncology Department at the Academic Hospital of Vrije Universiteit Brussel (Universitair Ziekenhuis Brussel, UZB), where he is also head of the Day Clinic in the Oncology center. As a member of the College for Oncology, he is in charge of setting out the national guidelines for the diagnosis and treatment of cancer. His clinical activities are undertaken in collaboration with several research laboratories. De Grève is (co-)author of more than 400 scientific articles and a member of the European Organization for Research and Treatment of Cancer (EORTC), the Belgische Vereniging van Interne Geneeskunde, the Belgian Society of Medical Oncology, the Belgische Vereniging voor de Studie van Kanker, the European Society of Medical Oncology, the European Association for Cancer Research, the American Association for Cancer Research, the American Society of Medical Oncology and the Scientific Counsel of the Interuniversity Course on Medical Oncology.

BREAKING THE DEADLOCK OF BUDGETARY AUTISM: WHAT ARE THE PARADIGMS FOR FUTURE HEALTH CARE ORGANIZATION IN BELGIUM?

MARC DE VOS & BRIEUC VAN DAMME[1]

At the beginning of the 21st century demographic, scientific and technological evolutions are increasingly putting financial strain on health care systems all over Europe, indeed in almost all developed countries. These evolutions are destined to increase as the century progresses, forcing governments, administrators, and health care professionals to think anew about the foundations of health care organization. In Belgium, the elementary pillars of our health care philosophy – quality combined with accessibility and free choice – are already eroding. A proactive and ambitious reform involving patients, providers, payers, the industries, policy makers, and academics will be needed to prevent further gradual decline[2].

Health care reform is not on the cards in Belgium today. The policy emphasis has been and still remains essentially budgetary. Therefore, for a coherent policy approach to be developed, we must first identify trends and challenges. Based on these a suggestion of possible policy options will be made. A pragmatic and realistic approach – we do not have the luxury of ideology or romanticism – can only be taken seriously if the priorities and limits of promising solutions are defined. The purpose of this article is to offer some food for thought on real health care policy reform in Belgium, based on the

---

1   Marc De Vos (Lic., LLM, Phd) is the Director of the Itinera Institute and a Professor of labor and employment law at Ghent University. Brieuc Van Damme (MA) is a fellow at the Itinera Institute and an independent consultant. The Itinera Institute is an independent think-tank and do-tank for sustainable economic growth and social protection, for Belgium and its regions: www.itinerainstitute.org. The authors thank Ivan Van de Cloot, chief economist of the Itinera Institute, for his input and constructive remarks. However, the opinions expressed in this article engage only the authors.
2   Daue, F. and Crainich, D., (2008). *Hoe Gezond is de Gezondheidszorg in België?*, Itinera Institute Report, online: www.itinerainstitute.org/upl/1/default/doc/20080421_SWOT%20 Deel%201%20NL_FVH_0.6.pdf.

stated necessity of such reform. Our purpose is not to provide a comprehensive or academic analysis, but rather to indicate – with a bird's eye view on the big picture – the unmistakable trends and future challenges that are facing us and to draw the plain conclusions they suggest.

Major trends in Belgian health care provision

1. Budgetary explosion combined with budgetary austerity

In health care organization – contrary to perhaps some popular and naïve belief about 'free' and accessible health care – everything comes down to numbers. And the numbers are impressive when you take a look at the evolution of the budget for public health care in Belgium. In 1970, public health care expenditure was still under the billion euro mark. Ten years later, it amounted to more than €3 billion. By the end of the millennium, public health care expenditure had reached €12 billion and it is very likely that this figure will again be doubled by 2010. In 2005, the public health care budget already equaled €17,25 billion, in 2006 €17,73 billion and in 2007 €18,87 billion. The objective is to spend approximately €21,5 billion in 2008[3] and €23 billion in 2009. Compare this figure with the €850 million in 1970 and the metaphor with the universe seems straightforward: always expanding and expanding. Of course, these are absolute figures. We have seen in over 30 years an average annual growth of close to five percent in real terms, i.e. on top of inflation. This is much faster than average economic growth in this country. From the perspective of public budget control, therefore, the growth of health care spending is simply unsustainable.

We have nonetheless managed to survive such an expenditure explosion by giving ever-increased weight to the relative importance of health care in the total social security budget. In 2008, the share of public health care expenditure in the total social security budget will be close to 32%[4]. In 1980, it was a mere 22%.

---

3   RIZIV/INAMI, (2007). *Statistieken van de geneeskundige verzorging*, online: www.inami.fgov.be/information/nl/statistics/health/2007/pdf/statisticshealth2007all.pdf.
4   This calculus under 'social security' also includes the public sector pensions, early retirements, and other social expenditure. From a more restrictive perspective, the share of health care is thus even bigger.

It is therefore fair to say that health care is gradually cannibalizing social security[5]. The victims of this budgetary evolution are the first pillar pensions, unemployment benefits and child allowances, all of which have seen their relative levels reduced because of increased health care expenditure. This situation is untenable in the long run and has already led to a series of health care policies that are perhaps necessary or inevitable, but that share a common characteristic in that they restrict the offer of, or access to health care in this country.

2. Health care policy vs. budgetary policy

Given the enormous and ever increasing budgetary importance of health care, it is normal and predictable that government should impose a budgetary discipline to avoid deficit spending. This necessary awareness, however, has turned into somewhat of an obsession. Since about a quarter of a century, Belgium's governmental policies in health care have indeed been dominated by budgetary concerns, rather than by public health concerns.[6] When one looks at the picture from a distance, one can easily come to the conclusion that health care policy in Belgium has essentially become budgetary policy. On the one hand, a lot of time and effort is spent on an almost yearly basis in determining growth norms for the public health care budget. On the other hand a number of reform measures, although not directly of a budgetary nature, have been developed under the growing pressure of budgetary austerity. In just the past couple of years, we have observed tightened budgets for hospitals and new technology, mergers of hospitals, and the concentration of some medical services in certain hospitals. The doctors and other health care providers have seen their therapeutic freedom restricted for the sake of efficiency. The freedom of choice in access to doctors is partially eroding and the doctors have seen the prescription of generic drugs imposed. More bureaucratic rules streamline the medical profession, the inflow of new medicines has been more strictly managed, reimbursed care is increasingly controlled, the inflow of doctors managed, etc.

---

5   Studiecommissie voor de Vergrijzing, (2008). *Jaarlijks Verslag*, online: http://docufin.fgov.be/intersalgnl/hrfcsf/adviezen/PDF/vergrijzing_2008_06.pdf. Studiecommissie voor de Vergrijzing, (2002). Jaarlijks Verslag, online: http://www.plan.be/admin/uploaded/200605091448049.OPVERG200201fr.pdf.
6   F. Daue and D. Crainich, *Hoe gezond is de Belgische gezondheidszorg?*, supra, note 3.

What the neutral observer notices, therefore, is a gradual streamlining and soft restriction on health care supply, and a gradual streamlining and soft restriction on health care demand. Some of these evolutions are highly contentious and debated. Many, indeed perhaps even all, may be necessary or desirable from a public governance perspective. But it goes without saying that they all have trade-offs. Our oft-trumpeted model of freely available and accessible health care in an open market that guarantees competition and choice is gradually eroding. What are widely considered as key components of the Belgian health care 'model' are thus being gradually undermined. We can illustrate this trend by focusing on two key parties: the medical profession and the patients/citizens.

3. The medical and paramedical profession under pressure

As almost any practitioner will tell you when questioned upon the state of his/her profession, doctors are facing less therapeutic freedom and more bureaucracy. Moreover, as hospitals have been rationalized, fewer have remained in the non-private sector, thereby decreasing the personal social security of the affiliated medical corps as compared to the previous generation doctors with public servant status. Furthermore, the income growth of the medical profession has diminished in relative size: between 1996 and 2008, the share of doctors' honoraria in the public budget went from 33,6% to 28,4% – a five points decline[7]. The systematic under funding of hospitals further exacerbates this trend, and has led hospitals to increase the overhead deducted from doctors' fees.

The growing pressures on the medical profession and its correspondingly diminished attractiveness should be a source of grave concern. Because at the end of the day, the quality of a health care system depends on the quality of its human capital. This goes for the medical profession as well as for the paramedical profession. Human resources will be a key challenge for the future wellbeing of Belgium's health care system. If we are to continue to thrive, we need to be able to attract and motivate the requisite human capital at home and, increasingly, abroad as well.

---

7   RIZIV/INAMI, (2007). *Statistieken van de geneeskundige verzorging*, online: www.inami.fgov.be/information/nl/statistics/health/2007/pdf/statisticshealth2007all.pdf.
    RIZIV/INAMI, (1999). *Statistieken van de geneeskundige verzorging*, online: www.riziv.fgov.be/information/nl/statistics/health/1999/pdf/statisticshealth1999.pdf.

4. Private expenditures are on the rise

Although our health care expenditures are financed by an ever-expanding public budget, the patients themselves have to carry some of the burden. The OECD computed that 27,7% of the total health care expenditures in Belgium are paid by the patient/citizen (or his/her employer), either as out of the pocket expenses or through private insurance[8]. Only four OECD countries have an even more important share of private expenditure: the US, Canada, Spain and Switzerland (Figure 1).

Figure 1: Health expenditure per capita, public and private, 2005
Source: Health at a Glance 2007, OECD Indicators.

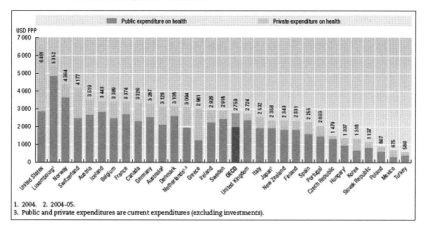

This already considerable share of private expenditure has been growing over the past few years, as can be seen from Figure 2 on the next page.

---

8   OESO, (2007). *Health at a Glance 2007*, OECD Indicators.

Figure 2: Evolution of the private share of health care expenditure in Belgium
Source: Health at a Glance 2007, OECD Indicators.

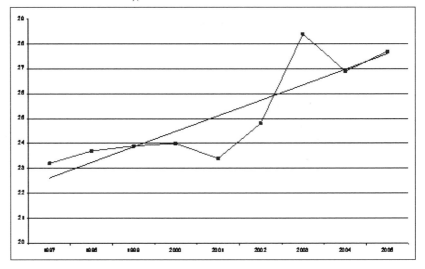

What is more, the share of private expenditure is likely to continue to increase in the years to come. Of course, this in itself is not necessarily a problem. Research shows that, while the marginal utility of consumption goods decreases rapidly with the number of purchases, this is not true for health care expenditure. In fact, "*as people get richer and consumption rises, the marginal utility of consumption falls rapidly. Spending on health to extend life allows individuals to purchase additional periods of utility. The marginal utility of life extension does not decline*"[9]. In other words, people are willing to pay for health care and for a whole bunch of health related goods and services, simply because they value them. This is one of the reasons why an increasing number of people are willing to pay for private health insurance.

---

9   Hall, R. and Jones, C., (2007). *The Value of Life and the Rise in Health Spending*, The Quarterly Journal of Economics, Vol. 122, nr. 1, p. 39 – 72, online: www.mitpressjournals.org/doi/pdf/10.1162/qjec.122.1.39.

According to the European insurance and reinsurance federation, the amount of privately insured individuals in Belgium has almost doubled in ten years: from 2,667 thousand in 1996 to 4,913 thousand in 2006 [10].

The growth of private expenditure signals the evolution of our economy towards a health economy. This is in itself a good and desirable thing and heralds the next phase in our economic development. On the other hand, however, and this is where our reflection kicks in, the growing share of private expenditure underscores the growing inability for public sector funding to match private health care demand. It underscores that the 'one size fits all' approach to health care provision fits less and less. As we shall see, this trend is not going to disappear and therefore policy debate should confront fully and squarely the question of choice and limits in public funding. The alternative is a continued slow erosion of publicly funded health care, with an American style proliferation of private insurance in a chaotic market context on the side. All this will be to the detriment of the poorest and sickest and is therefore not an attractive perspective and, we venture to claim, not a perspective the public would support if fully informed of the choice we face.

The budgetary challenge: How the problem can become (part of) the solution

Health care's place in society will be increasingly predominant in the 21st century, not only because of well-known demographic developments, but also because of socioeconomic, scientific and technological changes. The key challenge will increasingly be to provide health care that is both affordable and accessible, while being of high quality. The trends highlighted in the previous paragraph are therefore worrying. If we do not succeed in reversing them, these trends risk becoming real and structural weaknesses as the following decades unfold.

---

10 CEA Insurers of Europe, (2008). *The European Health Insurance Market in 2006*, CEA Statistics, nr. 35, online: www.cea.assur.org/uploads/DocumentsLibrary/documents/1218202930_european-health-insurance-2006.pdf.

1. The challenge of ageing

"In almost every country, the proportion of people aged over 60 years is growing faster than any other age group, as a result of both longer life expectancy and declining fertility rates. This population ageing can be seen as a success story for public health policies and for socioeconomic development, but it also challenges society to adapt, in order to maximize the health and functional capacity of older people as well as their social participation and security."[11] As we progress through the 21st century, global ageing will put increased economic and social demands on many countries. Belgium is no exception. As can be seen from Figure 3 below, the dependency ratio of the elderly compared to the population of working age is about to double in 50 years. In 2050, there will be 2,27 people of working age, for one elder (65+), which is about half of the ratio at the end of the 20th century.

The dramatic decline in the number of (potentially) economically active people as compared to the number of (potentially) economically inactive is of major concern in countries where – as in Belgium – social security (including health care) is financed through the so called 'repartition' system. In such a system, the social security expenses for the old and the sick are paid for by the current working generation, who themselves hope that the future generation will do them the same favor. However, with the ageing and retirement of Baby Boomers, the equilibrium between succeeding generations disappears and our society is consequently faced with a real and inevitable budgetary challenge. Needless to say this is going to put tremendous pressure on our social security system and thus on taxpayers' contributions.

Moreover, the ageing of the population as such is also estimated to increase health care expenses by three percentage points by 2049, as can be seen from Figure 4. Roughly, this represents €10 billion more expenses. This rise in expenditure comes on top of the dramatic doubling of the dependency ratio. Fewer and fewer younger workers will have to finance ever-increasing health care expenses for more and more older retirees.

---

11   Gro Harlem Brundtland, Director-General, World Health Organization, 1999, cited in WHO, (2002). *Active Ageing, a policy framework, A contribution of the World Health Organization to the Second United Nations World Assembly on Ageing*, Madrid, Spain, April 2002, online: http://whqlibdoc.who.int/hq/2002/WHO_NMH_NPH_02.8.pdf.

Figure 3: Dependency ratio of the elder
(Ratio of the population of 65 years or older on the population of working age)
Source: National Bank of Belgium [12]

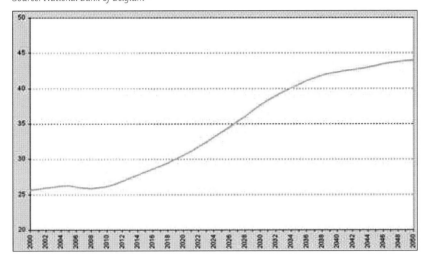

Figure 4: The budgetary cost of ageing
(Percentage points change of GNP, compared to 2007)
Source: National Bank of Belgium

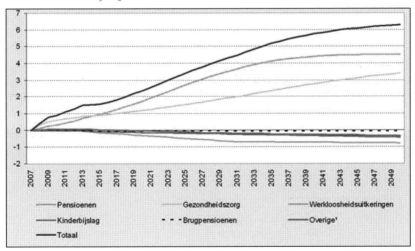

---

12   Presented by Jan Smets, Director of the NBB, on the CEDER study day 'Aan de slag (blijven)', 05/09/2008.

2. More health and the rise of the health economy

For Belgium, it is estimated that the phenomenon of ageing in itself will 'only' increase health care expenses by 0,7% on an annual basis [13]. The expected larger share of health care in our economy can therefore not be explained by demographic factors alone. Ageing is just the tip of the iceberg of growing health and health care expenditure. It is widely acknowledged that several drivers will be responsible for an inexorable push in health care expenditures in the decades to come, besides demographics [14]:

- Changing lifestyles and the consequent explosion of lifestyle diseases, e.g. related to obesity.
- Continued increased specialization in the medical profession, as the scientific evolution creates ever more avenues and branches.
- Innovation in technology and medicines, opening up new treatments and narrowing down the target group to eventually the level of individual and genetic treatment, where the cost saving effects of blockbuster treatments with huge markets will disappear. The treatments will continue to improve, but their relative cost will rise.
- Consumerism, as people become ever more demanding and willing to improve their health and wellbeing, further blurring the line between medicine and consumption.
- Greater wealth in both the Western world and the now rapidly expanding developing world, further feeding the desire and willingness to pay for health and health care.

The link between 'wealth' and 'health' in the shape of health care expenditure is borne out by economic research, also in Belgium. The Federal Plan Bureau found that the 'elasticity' of health expenses per capita and GDP per capita – this is the extent to which health expenditure reacts to increased economic growth – is superior to one [15]. This means two things: (1) that the wealthier people become, the more they are willing to spend on health care,

---

13 Van de Cloot, I., (2003). *De Beheersbare Gezondheidszorg, Financiële Berichten ING*, Nr. 2390, p. 1 – 10; NBB, Jaarverslag 2003, p. 94-97. www.nbb.be/NR/rdonlyres/9C708875-6591-41C2-AFF0-2E40BC1E1F33/0/JV2003T1_volledig.pdf.

14 See, inter alia, Boston Consulting Croup, (2007). *Health Care Regulation Across Europe, From Funding Crisis to Productivity Imperative*, online: www.bcg.com/impact_expertise/publications/files/HealthCare_Regulation_Europe_Sept_2007.pdf.

15 Studiecommissie voor de Vergrijzing, (2002). Jaarlijks Verslag, online: www.plan.be/admin/uploaded/200605091448049.OPVERG200201fr.pdf.

and (2) people are prepared to spend proportionally more on health compared to the extra wealth they have acquired. The relationship between GDP per capita and health expenditures is also illustrated in Figure 5 below: the 'wealthier' a country, the 'healthier' a country.

Figure 5: Health expenses per capita and GDP per capita, 2005
Source: Health at a Glance 2007, OECD

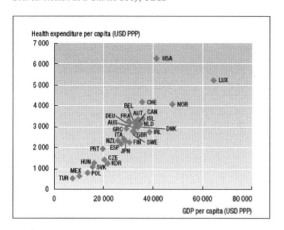

All of this indicates that the citizen/patient is consciously choosing for health. Citizens are no longer mere patients who swallow whatever the doctor prescribes. They are becoming more and more conscious health care buyers and consumers, further stimulated by increased access to health care information via a variety of sources, including the Internet. According to some long-term estimates, up to one third of a developed country's GDP will thus be spent on health care by the end of our century[16]. *This signals the evolution of our economy towards a health economy, a new stage in economic progress in post-industrial societies.*

Economically speaking, it makes no sense to deprive people of something that creates value for them. The bottom line is: rather than being satisfied with the landscape as we know it now, which is characterized by ever more rationing, more trade-offs and more multi-speed medicine under an ever tighter public budget, we should allow ourselves to invest more and more

---

16 Getzen, T., (2008). *Modeling Long Term Healthcare Cost Trends, Research Projects in Health, the Society of Actuaries,* online: www.soa.org/research/health/research-hlthcare-trends.aspx.

consciously in health care. *We should put our traditional budgetary autism aside and start grasping the economic opportunities that arise from the shift to a health economy.* In this perspective, the growing private expenditure on health care is not so much the problem as it is part of the solution. For that to be the case, however, the necessary but narrow focus on budgetary control in public sector funding needs to be lifted and an open debate on the limits and choices in public health care provision must be recognized as both inevitable and desirable.

The necessary quest for new paradigms for future health care organization in Belgium

1. Moving beyond the 'Brussels consensus'

To anticipate the budgetary impact of ageing and to discipline governments in the short run for this long-term challenge, Belgium has established a virtual savings strategy (the Silver Fund) and an annual rite of future ageing cost estimation, in the shape of reports from the High Council of Finance's Commission on Ageing. The purpose of these reports is to estimate – and the estimates have never ceased to increase with each annual report – the expected future cost of demographic ageing, *based on a number of parameters of future fiscal, social, and economic performance.*

These parameters can be summed up as follows. We contrast the premise with the past/current performance:

- 1,75% labor productivity growth per year until 2030 (average 1,45% between 1980-2005).
- Total unemployment rate of 8% in 2030 (12,6% in 2007).
- Activity rate of 70% in 2030 (62% in 2008, or a difference of roughly 500.000 jobs).
- Average annual economic growth 2,2% until 2030 (1,8% between 1990-2005).
- State debt 60% GDP in 2014 (81,4% in 2008).
- Annual average real-term growth of the public health care budget restricted to 3% until 2030 (near 5% between 1970-2006).

As the list shows, the estimates assume a systematic and marked improvement of Belgium's fiscal, social, and economic performance. We have argued elsewhere that such an improvement is very unlikely and in fact amounts to wishful thinking without prior fundamental policy reform[17]. Indeed, *ceteris paribus*, population ageing is likely to impede and slow economic performance, not improve it[18]. The High Council of Finance itself recognizes the limits of a purely budgetary strategy and advocates reforms that stimulate growth and employment in order to meet the financial challenge of ageing[19].

More importantly for our exercise is the last of the aforementioned premises, which seeks to reduce the annual growth rate of public expenditure on health care to 3% per year, i.e. a baffling reduction of almost 40% as compared to the average growth in the previous 35 years. Given the powerful vectors that will increase rather than decrease health care needs in the future, as listed above, this estimate is simply unbelievable. In the absence of fundamental reform in both health care organization *and* in health care financing it can only mean a growing 'sovietization' of Belgian health care for the general public, with an increasingly important private market for the fortunate. This is a proposition too undesirable even to entertain.

Given these stark realities, the Belgian health care system – both in its organization and financing – has no option but to reform and improve. Without such reforms we will simply not be able to maintain anything near the quality and accessibility we now enjoy. This article is not the place for developing a comprehensive and balanced list of fundamental proposals.[20] We will, however, indicate some directions with potential, in the hope of broadening mindsets and starting a pragmatic debate on the paradigms of future health care organization in this country.

---

17 De Vos, M., (2008). *Doorbreek de cijferban van de vergrijzing*, Itinera Institute Nota, online: www.itinerainstitute.org/upl/1/default/doc/20080708%20-%20Nota%2028%20-%20 Doorbreek%20de%20cijferban%20van%20de%20vergrijzing%20-%20MDV.pdf
18 Gruescu, S., (2007). *Population Ageing and Economic Growth*, Contributions to Economics, Springer publishing.
19 In Federal Public Service Finance, (2008). *Belgium's Stability Programme (2008-2011)*, online: http://stabiliteitsprogramma.be/en/ Stabilityprogramme_2008_2011_Belgium_Cabinet_Finances_20080418_EN.pdf.
20 The interested reader can find food for thought on these in F. Daue and D. Crainich, *Hoe gezond is de Belgische gezondheidszorg*, see note 3 above.

2. The promise of ICT

Information and communication technologies (ICT) have already had a significant impact on economic growth, but also on health care and the delivery of health services in a number of countries. From telemedicine to electronic health records to RFID[21] to embedded sensors, a variety of health ICTs have been shown to improve operational and administrative efficiencies, clinical outcomes, documentation and information flow in a variety of global settings. Chaudhry et al. (2006)[22] have scrutinized 257 empirical studies to analyze the impact of health information technology on quality, efficiency and costs of medical care. The analyzed studies unanimously reported positive results on the quality of care through an increasing adherence to guideline or protocol-based care, clinical monitoring based on large-scale screening and aggregation, transparency, and the reduction of medical errors. ICT was also found to improve health care efficiency thanks to more accurate diagnosis and thus less unnecessary treatments and medication consumption. One examined study reported efficiency gains up to no less than 24%. Chaudry et al. were not able to find relevant studies – they were either too old or methodologically questionable – that showed ICT to be cost reducing in health care.

On the other hand, Hillestad et al. (2005)[23] computed a cautious estimate – not a proof – of how much money could be saved in the US thanks to the generalized application of the electronic health record[24]. The estimation yielded an impressive figure of $513 billion by 2020.

What the above demonstrates and illustrates is the potential of information and communication technology to improve the organization of health care, to improve the delivery of health care services, to improve health outcomes and to rationalize health care spending without restricting the supply of

---

21 Radio Frequency Identification is an automatic identification method, relying on storing and remotely retrieving data using devices called RFID tags or transponders.
22 Chaudry, P. et al., (2006). *Systematic Review: Impact of Health Information Technology on Quality, Efficiency, and Costs of Medical Care*, Annals of Internal Medicine, Vol. 144, Nr. 10, p. E12 – E22.
23 Hillestad, R., et al., (2005). *Can electronic medical record systems transform health care? Potential health benefits, savings, and costs*, Health Affairs, Vol. 24, nr. 5, p. 1103 – 1117.
24 An electronic health record (EHR) refers to an individual patient's medical record in digital format. Electronic health record systems coordinate the storage and retrieval of individual records with the aid of computers.

health care services. In view of the current pressures and future challenges facing the Belgian health care system, it is clear that these benefits represent both an opportunity and a necessity. ICT should be and will be central to the future of Belgian health care organization, much more so than it is today and than current government programs envisage.

3. Horizontal v. vertical integration of health care services

The Belgian health care system is essentially *vertically integrated*. From the top down, the government decides on budgets, the RIZIV/INAMI allocates budgets, the mutual funds (or private insurers) assure reimbursement, the hospitals organize and centralize care, the specialists provide specialist care, and the general practitioners provide general care. This slicing up of the health care cake induces turf wars and causes mutual isolation between different levels in health care provision. From the perspective of health outcomes this is a suboptimal situation, especially since a large percentage of health care expenditure is linked to a limited group of pathologies. It would be more logical, and indeed more productive, to adopt a *horizontal approach* where the main pathologies would be targeted in a succession of stages: from information and sensitization (prevention), to screening, early diagnosis, and eventually team treatment with various health care professionals involved in the particular disease on a platform basis. Health care providers, with the right government support and structure, could thus work more closely together to improve coordination and access to health care, and to ensure better health outcomes. Today's parceled out approach could make room for a *continuum of care* which integrates the whole health care chain.

According to the World Health Organization, the continuum of care offers a complete service array, from hospital to home care, and requires all medical and social services within the community to be brought together. The connection of all health care initiatives on all levels of the health care system is also part of the continuum of care. The patient therefore stands at the centre of the health supply chain. For every patient and for every type of pathology, the most adequate and available treatment is suggested. Not the profitability for any one level or actor, but patient's needs are the most important selection criterion when treatment is offered. Obviously, this implies more coordination and integration between the different health care levels and health care services.

The distinction between a 'vertical' and a 'horizontal' approach to health care is not sacrosanct. There are, for instance, certainly issues of organizational complexity in framing a horizontal, disease and patient oriented approach. But what the above illustrates is the need for the Belgian health care organization to reconsider both the individual role of the respective levels or actors in health care organization and the way they collaborate for ensuring optimal health outcomes with improved efficiency. Is the division between GPs and specialists useful? What role do mutual funds have to assume going forward? Should not the patient and the disease be central to the process, rather than the institutional structure of health care? The current vertical division of health care organization does not easily allow such reconsiderations, but on the contrary reinforces conservative and interest group style reflections (soft corporatism) at the expense of efficiency or health outcome. We need the freedom to reconsider the relevance and purpose of the current institutional actors in the health care system if we are to preserve its health care performance for the future.

4. Towards a real debate on a multiple pillar structure in health care

We have seen that:

- While even today a large percentage of Belgian health care expenditure is already private;
- Public health care expenditure in the future will increasingly suffer from the gulf between what is required and what is affordable, as the Belgian repartition system meets the combined challenge of ageing and the exponential growth of health care demand.

This sober reality should force us to recognize what is already a reality today and what will increasingly become a necessity tomorrow, i.e. that health care funding is both a public and a private affair. The solid policy approach is not to deny this combination but to confront it and have a societal debate about the combination and organization of both. The policy of denial, which is often practiced today, offers no respite but instead allows private funding to develop organically in an unregulated market. This results in limited transparency, unlimited price increases, and a real two-speed society between those who can and those who cannot afford private insurance of some kind. If you are looking for the USA, you need not look any further.

The very sensitive debate about the limits of public health care provision needs to be brought into the open. It is currently hidden behind the closed doors of administration and a mass of ad hoc decisions on public funding. It will, of course, be a very difficult and sensitive debate. The limits of public health care provision will have to be determined, not on ad hoc basis but on a fundamental and principled societal basis. The role and responsibility of various actors will have to be (re)defined, since we would have to organize additional pillars of health care funding by recalibrating the responsibilities of citizens, employers, insurers and mutual funds. In the same vein, patient responsibility would have to be constructed and organized, implying a variety of ethical questions on the limits of solidarity and the scope of personal responsibility.

The debate will thus undoubtedly be difficult, but at the same time cathartic. It will allow us to rationalize and democratize the vagaries of currently ad hoc budgetary decisions. It will allow us to streamline and organize a market for private insurance, ensuring due attention to coverage of the poor and the ill. It will allow us to set ethical rules of personal conduct and responsibility, making the residual solidarity fairer and more defensible. And, as emphasized above, it will allow us to liberate the funds necessary for our inexorable and fortunate evolution towards a health care economy. The alternative is political meandering, ethical distortion, and budgetary scarcity. Multiple pillars of health care financing will be inevitable and necessary if Belgium wants to maintain, not only a high level health care system but also a fair and just health care system.

Conclusion

The traditional public rhetoric leads Belgians to believe that theirs is one of the best health care systems in the world. The accolade may or may not be true. What is certainly false, however, is the common political conclusion that the only debate should be about how much public money is poured into the system. This political mantra, which has dominated Belgian health care policy for the past quarter of a century is untenable if we are to successfully confront the twin challenges of ageing and increased health care demand in the 21st century. These challenges will be inevitably upon us for the coming decades. How can we meet them while maintaining the real *fortes* of the Belgian health care system, i.e. quality and accessibility?

This short paper argues that we will certainly not meet the impending challenges if we follow the wholly unrealistic 'Brussels Consensus' on the impact of ageing. We will find ourselves in a very uncomfortable dead end street if we do not succeed in adopting reform policies that improve both health care funding and its performance. The foundations of the Belgian health care model – quality combined with accessibility and choice – are already gradually eroding. Only by considering new avenues for its organization and financing will we be able to sustain for future generations the type of health care performance we enjoy today.

We suggest three lines of thinking: increasing investments in ICT, improving coordination and integration between the stakeholders of the health care system, and a real debate on a multi-pillar structure for the financing of health care. These are nothing more than openings for debate. The question is whether the political and institutional health care community in this country, which is so mobilized by the day-to-day constraints and challenges, will be able to entertain creative and fundamental thinking in time. A health care system is like the proverbial tanker that turns ever so slowly but which consequently is equally hard to correct once it has turned. Let us hope, for all our sakes, that the Belgian actors will turn in time.

BIOS

Marc De Vos holds a Licentiate and Doctorate in Law (Ghent University), a Master in Social Law from the Université Libre de Bruxelles (ULB), and a Master of Laws (Harvard University). He is director of the Itinera Institute and teaches Belgian, European and international employment and labor law, as well as American law, at Ghent University and the Vrije Universiteit Brussel (VUB). He frequently publishes, lectures and debates on issues of labor and employment law, European integration, labor market reform, pensions, health care, ageing and the welfare state, both nationally and internationally, and in academic, professional and policy circles, as well as in the media.

Brieuc Van Damme holds a Bachelor Degree in Economic and Social Sciences (Namur Universisty, 2004), a Master in Economic Sciences – international and development economics specialization (Leuven University, 2006) and a Master of Arts in European Economics (College of Europe, 2007). After his Government and Public Services consulting job at Accenture, Brieuc joined the Itinera team in March 2008.

PART TWO
SHARED RESPONSIBILITY

WILLINGNESS TO PAY AND SOLIDARITY

ERIK SCHOKKAERT

When listening to political discussions in Belgium, one might get the impression that a good Minister of Social Affairs should simply keep health care expenses as low as possible. Yet, this is obviously too simplistic, since there are not only health care expenses but also health benefits. A zero level of expenses is far from optimal. This raises the question: is there a socially desirable level of expenses in health care? To make this question operational, economists developed the concept of 'willingness to pay'. A lot of people have an allergic reaction to the mere idea of 'willingness to pay' in health care since they state that something as holy as 'life' can and never should be expressed in terms of money. Yet, as I will argue throughout the rest of my article, this reaction rests on a misconception.

On the following pages I will reflect in a rather abstract way on the notion of 'willingness to pay' and on the relationship between willingness to pay and solidarity. I focus on the question: how to think about an 'acceptable' level of health care expenses? I will not discuss in depth the problem of the efficiency of the health care system itself. In every insurance system, there are problems of what is called 'moral hazard' and health insurance is no exception. It is clear that these efficiency issues are linked to the question of the optimal size of the health care sector. Surely, the willingness to pay of the population will be influenced by their perception of (in)efficiency. However, I will not focus on this point, because I take it for granted that good institutions should aim at avoid overconsumption.

Global approach

My starting point is the sharp rise in public health care expenditure. In recent decades, RIZIV-INAMI[1] expenditure in our country has been steadily rising. It increased from about €7 billion in 1990 to more than €18 billion in 2006. This increase in expenses has led to questions about its justification. Is the increase desirable and sustainable? In fact, in almost all Western countries, reactions have been rather similar[2]. Whilst after World War II there was initially much enthusiasm about the development of universal systems of health care, by the 1980s there was growing concern about the increase in expenditure and a growing desire to control it. At that stage, general budgetary restrictions and price-regulations were implemented. From the mid-1990s onwards, different countries entered a third phase in which gradually more and more micro economic financial incentives were introduced. This general pattern of reactions begs the real question: is the growth of health care expenditure indeed problematic?

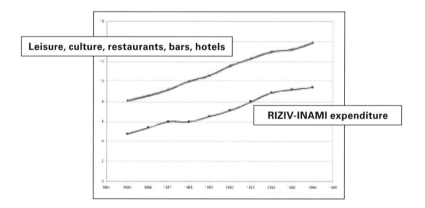

The graph above clarifies the question. It represents the evolution in Belgium between 1985 and 1995 of two categories of spending. The line below shows RIZIV expenditure, the upper line shows expenditure on leisure, cul-

---

1   RIZIV-INAMI or *Rijksinstituut voor ziekte- en invaliditeitsverzekering/Institut national d'assurance maladie-invalidité* is the Belgian State institute for sickness and invalidity insurance.
2   Cutler, D. 2002. *Equality, efficiency, and market fundamentals: the dynamics of international medical-care reform*. Journal of Economic Literature 40: 881-906.

ture, restaurants, bars and hotels. I focus on the decade 1985-1995, because this was the period in which worries about health expenditure became serious. But why should the rise in health care expenses be seen as problematic, while leisure expenditure is apparently not perceived as such? Surely, health is as important as recreation? I suggest that there are two possible hypotheses: (1) health care expenditure is not decided by the market and political decisions do not reflect the true willingness of citizens to pay; and (2) health expenditure is financed in a collective and redistributive way, and this imposed solidarity may have come under pressure.

The first hypothesis is not very convincing. Some interesting (mostly US) studies have appeared recently, arguing convincingly that the willingness to pay for benefits in health is in principle very high. The most basic argument is formulated by Hall and Jones as follows: *"(...) as we get older and richer, which is more valuable: a third car, yet another television, more clothing – or an extra year of life?"*[3] The answer is obvious and does not require sophisticated economic insights: most people would prefer to live another year. A second argument points to the complementarity between different forms of technical progress in health care: *"Improvements in life expectancy raise willingness to pay for further health improvements by increasing the value of remaining life. This means that advances against one disease, say heart disease, raise the value of progress against other age-related ailments such as cancer or Alzheimer's."*[4] To state it simply: if you know that Alzheimer disease can be better treated, it becomes more interesting to live longer.

The same authors also propose a simple formal model to think about how much a nation should spend on health care, and what is the time path of optimal health spending. The basic idea is that with limited resources, society has to trade-off material wellbeing against better health. The optimal level of health spending will then be determined by (1) the relative importance attached to consumption versus living longer and in better health, and (2) the efficiency of health care expenditure in influencing life expectancy and life quality. I noted already that if society grows richer, the relative value of consumption decreases. Combining this insight with the best empirical

---

[3] Hall, R. and C. Jones. 2007. *The value of life and the rise in health spending.* Quarterly Journal of Economics 112: 39-72.
[4] Murphy, K. and R. Topel. 2006. *The value of health and longevity.* Journal of Political Economy 114: 871-904.

material/data on the efficiency of health care spending that are available for the US, Hall and Jones find that socially optimal expenses in health care will strongly increase in the following decades and could by 2050 represent about 35% of the gross domestic product. Sensitivity analysis shows that with a larger (but still reasonable) value for the quality of life, one could even end up with an optimal share of 45%. Similar results have been found for other countries. Note that this does not mean that material prosperity has to decline since the studies suppose an economic growth of 2 to 3% per year. Health and material prosperity will both increase. All this strongly suggests that the willingness to pay for health care is high and that from a social welfare point of view it would be optimal to allow health care spending to increase in the future. Apparently it is not unwillingness to pay for health care that would be the cause of growing concern about the increase in health care spending. Let me therefore now turn to the second hypothesis: the pressure of solidarity.

Willingness to pay and solidarity

In the market, everybody pays for their own consumption, e.g. restaurants or travel. The same is basically true in private insurance markets where premiums are related to the degree of risk involved. Our collective health insurance system, however, imposes solidarity between good and bad risks. The young and healthy do not pay a lower premium than the old and the sick. And with contributions related to income, there is in addition some solidarity between higher and lower incomes. The main question is: is this (imposed) solidarity still supported by the population?

In 2002, Marc Elchardus et al. conducted a representative survey in Flanders to measure the public support for the welfare state[5]. A majority of the respondents accepted the basic principle of social security. The insurance idea was particularly popular. There still remained a majority – although smaller – who accepted solidarity as a moral duty towards the weakest individuals in society. Elchardus et al. also asked about responsibility in health insurance. Almost nobody agreed with the statement that people with a genetic disposition to developing a serious illness later in life, should contrib-

---

5   Elchardus, M., A. Derks, M. Debusscher, and C. Tresignie. 2002. *Het draagvlak van de solidariteit*. VUB – TOR.

ute more. A slightly larger number of respondents (6%) thought that older people should contribute more. More than 11% would reward people with a healthy lifestyle by lowering their contributions [6].

These results sketch a picture of the situation at a given moment in time. More important for our question is the development of these feelings over time. There is growing evidence to show that feelings of solidarity have been declining in recent decades. Firstly, there is a trend towards what we could call a more fragmented society – more people behave as individual consumers, social distance increases, and altruism becomes less obvious [7]. Data from the World Values Survey [8] show that most people are prepared to do something to improve the circumstances of family members. Doing something for neighbors is less popular and helping immigrants even less so. Secondly, social norms with respect to financial incentives and personal responsibility are shifting. In the large majority of countries, the share of respondents agreeing with the principle that those who are more productive should be paid better, is increasing over time [9]. And it is well known that there is a strong correlation between support for the welfare state and the importance attached to individual responsibility [10].

Let me draw some provisional conclusions. First, at an aggregated level the willingness to pay for health care is high. Second, our system of health insurance imposes a large amount of solidarity and this solidarity is coming under more and more pressure in recent decades. The challenge is then clear. How to mobilize the existing willingness to pay in an equitable way, i.e. what public institutions should be set up to avoid that the distribution of health care in the future will be disproportionately in favor of the rich?

---

6   In fact, in other countries and with other samples, the percentage of respondents who want to make people responsible for lifestyle is much larger – for the US see, for example, Fuchs, V. 1996. Economics, values and health care reform. American Economic Review 86: 1-24.
7   See, for example, the different contributions in Van Parijs, P. 2004. *Cultural diversity versus economic solidarity*. Bruxelles: de Boeck.
8   The data from the World Values Survey can be downloaded from www.worldvaluessurvey.org/.
9   The wording of the question in the World Values Survey goes as follows: *"Imagine two secretaries of the same age, doing practically the same job. One finds out that the other earns considerably more than she does. The better paid secretary, however, is quicker, more efficient and more reliable at her job. In your opinion, is it fair that one secretary is paid more than the other?"*
10 See, for example, Alesina, A. and G.-M. Angeletos. 2005. *Fairness and redistribution*. American Economic Review 95: 960-980.

Economic laws?

Let me pause here and ask an intermediate question. Is it not the case that the economic constraints on decision-making are so stringent that we have no choice but to cut expenditure? The answer to this question is a definite 'no'. When we go beyond ideology, there is much more going on than the simplistic story of a socioeconomic model that is on the verge of collapse due to high social spending. To illustrate: De Grauwe and Polan [11] show that the higher the share of social spending in the GDP of a country, the more competitive it is. In fact, there are good economic explanations for this observation [12]. People want to insure themselves in order to avoid risks – and in many cases social insurance is more efficient than private insurance [13]. A well-educated and healthy population is more productive. The welfare state lowers insecurity and helps to avoid deep social conflicts. Lindert also suggests that most of the countries on the European continent adapted their tax systems to accommodate larger social expenditure, e.g. by giving more importance to indirect taxes. There are indeed many economic arguments to support the idea that a larger welfare state is not necessarily bad.

Neither do we have to exaggerate the 'automatic' effect of social contributions on labor costs. This is clearly illustrated by the US debate about mandated benefits [14]. Consider the effect of larger contributions on the process of wage negotiations. Suppose the workers are aware that larger contributions are linked to a better social protection, i.e. that the benefits they receive from the system are more or less proportional to their contributions. This is different from the perception of (progressive) tax payments, in which case there is no direct link between payments and public expenditure. Workers may realize that the alternatives to social insurance are either private insurance or private savings – both of which may be more costly. Certainly in a system of collective negotiations, trade unions could accept a lower net

---

11   De Grauwe, P. and M. Polan. 2005. *Globalisation and social spending. A race to the bottom?* Pacific Economic Review 10: 105-123.
12   Lindert, P. 2004. *Growing public.* Cambridge: Cambridge University Press.
13   See, e.g., Barr, N. 1998. *The economics of the welfare state.* Oxford: Oxford University Press and Schokkaert, E. and F. Spinnewyn. 1995. *Fundamenten van sociale zekerheid: solidariteit en verzekering, overheid en markten,* in Despontin, M., and M. Jegers (eds.) *De sociale zekerheid verzekerd?* (Brussel: VUBPress): 223-268.
14   A brief sketch of the issues involved is given by Summers, L. 1989. *Some simple economics of mandated benefits.* American Economic Review (Papers and Proceedings) 79: 177-183.

wage in return for better social protection. The problem of wage costs then boils down to an empirical question. Is there a link between contributions and benefits? The stronger the focus on insurance, the smaller the unfavorable job market effects [15]. And do workers take into account that link during wage negotiations? Note that the notion of willingness to pay plays a crucial role here. First, the labor market effects will depend on the willingness to pay. The more the benefits are perceived as benefits, the smaller the effect on wage costs will be. Second, the greater the willingness to pay, the more politically acceptable the system will be to the population.

All this brings us back to our main question. Is it possible to 'organize' the system in such a way that growing health care demands can be channeled in an equitable way? How to convince people that an increased share of public health care spending in GDP is socially desirable and may even be in their own personal interest? Of course, the trade-off between material welfare and health care spending will be more favorable if the system of health insurance becomes more efficient. As already mentioned, fighting moral hazard is essential. It is not sufficient, however. In my view it is equally crucial to increase the transparency of the decision-making process and to clearly establish for the population the trade-offs involved. We should perhaps not be over optimistic about the effects of this policy, but there seems to be no other way.

The role of transparency

Before turning to the issue of solidarity, we should first exploit the insurance idea as far as possible. Let me illustrate this with a simple example. It is well known that (in all countries) 5% of the population accounts for 50% or more of the health care expenses. One possible interpretation of these figures would be to appeal to (ex post) solidarity between the 95% of 'healthy' people and the 5% that are very ill. Yet one can also look at the figures in another way and stress the (ex ante) uncertainty. Take any group of people and we can predict that one out of twenty will sooner or later be confronted

---

15 In the case of pensions this has by now become the official position of the World Bank. For empirical results supporting the hypothesis, see Ooghe, E., E. Schokkaert, and J. Fléchet. 2003. *The incidence of social security contributions: an empirical analysis.* Empirica 30: 81-106, and Disney, R. 2004. *Are contributions to public pension programmes a tax on employment?* Economic Policy 19: 267-311.

with high health care expenses. Yet we do not know who this will be. It can be anyone of us – or, more broadly, it can be one of our children. This is precisely the justification for a health insurance system. Highlighting this insurance aspect is probably the most powerful argument to justify the present system.

Yet it will not be sufficient to safeguard a high degree of solidarity between bad and good risks and between rich and poor. Increasing transparency in this respect would draw attention to the trade-off between material consumption and health care – which brings us immediately back to the idea of willingness to pay. Is this transparency psychologically feasible? [16]

Psychologists have pointed out that health and life are so-called 'sacred values'. An extra car or material prosperity are 'vulgar values'. People have no problem with choosing between different vulgar values, to consider *routine* trade-offs. Every day we make choices about how to spend our money. People also accept *tragic* trade-offs, i.e. trade-offs between sacred values – think about Sophie's choice or about the Greek tragedies in which the hero has to choose between his nation and his life. Yet, we have serious difficulties with *taboo* trade-offs, with choosing between sacred and vulgar values. There are strong psychological mechanisms at work that make us oppose such choices. We try to ignore them. We refuse to see relevant statistical truths or to follow a logical line of thinking.

Choices in health care are exactly of this kind. How much material consumption should we give up to offer cancer patients the chance of surviving a few more months? Having to weigh up one life against another is hard. Weighing up material prosperity against health is scandalous. Yet, while we are reluctant to face such choices, we still have to make them. We then turn to all kinds of escape routes. We raise curtains. More or less secret expert commissions are installed, where the difficult choices are packaged as technical decisions. By focusing on monetary costs and not on health benefits, we disguise the taboo trade-offs as routine trade-offs. While psychologically understandable, there is a huge danger here that these decision-making procedures will make us lose sight of the real challenge lying before us. I pro-

---

16 A brief overview of the literature can be found in Tetlock, P. 2003. Thinking the unthinkable: sacred values and taboo cognitions. Trends in cognitive sciences 7: 320-324.

pose that we try to look the issue straight in the eye. Since we cannot escape making taboo trade-offs, we had better be open about them. We should try to make the trade-offs as transparent as possible. Let me give two specific examples: the issue of the growth norm and the economic evaluation of expensive medicines.

The growth norm

The introduction of a yearly growth norm for health care expenditure in Belgium has undoubtedly had some positive effects. The norm acts as a reference for government decisions and sharpens the concern for efficiency. However, my personal feeling is that the growth norm has decreased rather than increased transparency about the real issues in health insurance. Firstly, there is the implicit suggestion that it is only possible to keep to the norm if the government would become more 'efficient'. Public opinion then gets the overly simplified impression that the government can directly control expenditure. Yet health care spending reflects the day-to-day decisions taken by the actors in the field. It is not the Minister of Social Affairs who decides about drug prescriptions or about the specific medical interventions for patients. His decision space is much more limited than in other government domains. Secondly, and more importantly, the discussion about the growth norm has rarely been embedded in a broader social perspective. It is a good example of how to disguise taboo trade-offs as routine trade-offs between purely material values. One can even say that to some extent we have succumbed to number fetishism. If everybody keeps working under the illusion that simply improving efficiency will allow us to satisfy any growth norm it is no longer necessary to formulate and face the hard choices.

Let us therefore make explicit the consequences of a (too) low growth norm. It will unavoidably lead to a continuing (often implicit) privatization and to the erosion of coverage in the public health insurance system. Given the willingness to pay, health care spending will keep growing, but the growth will take place without the solidarity of the collective system. Moreover, systematically exceeding the growth norm also has psychological consequences. Panic stricken reactions by politicians and by the media immediately create the impression that the increase in health care spending is problematic, sometimes even that the whole organization of the system is problematic. Even if the growth norm was initially defined in an arbitrary way, exceeding it is still perceived as a sign of failure. These reactions have nothing to do with economically justified thinking about the optimal allocation of resources. They are very counterproductive in the light of the challenge that I am describing in this article.

In the light of this challenge, the debate about the growth norm should be broadened. Instead of focusing on a specific number, politicians and political parties should be invited to make explicit which treatments they want to include in the coverage of the compulsory system and which treatments they want to relegate to the private supplemental insurance system. They should make explicit how they want to deal with the expected increase in health care costs resulting from technological progress in the sector. Note that this is *not* a plea for a high growth norm; it is a plea for transparency! It is also a plea to avoid the acute hypocrisy that is now dominating the debate, with players making proposals whilst knowing that they are unfeasible.

Economic evaluation of expensive medicines

My second example is economic evaluation and the way we deal with expensive medicines. Economic evaluation of such expensive medicines is essential and cost-effectiveness is necessary to keep the system affordable. However, it is striking how the official institutions in charge of such evaluations (such as NICE in the UK or the KCE in Belgium) shy away from cost-benefit analysis, in which the willingness to pay is introduced explicitly. Basically, they shy away from making taboo trade-offs. Yet I suggest that the alternative evaluation methods based on cost-effectiveness that are most popular now, are not sufficient for making explicit the important trade-offs and for making the debate fully transparent.

A first problem is that distributional judgments are almost always neglected. Yet, the benefits of an intervention may be distributed unevenly over the population. And in many cases the desirability of taking up an intervention in the coverage of the compulsory system may depend on the way it will be financed, more specifically on the structure of the personal co-payments. Not making distributional judgments explicit does not mean that one avoids making them, only that one is not aware of their consequences.

A group of researchers surveyed 328 economic evaluation studies from the period 1995-1997. The results are devastating. Within these 328 studies, there is not a single one (!) that gives detailed information about the distributional consequences of the decision: *"The overall picture (...) is certainly the worst that could have been expected before starting this literature review. (...) Despite the significant and increasing number of economic evaluations published every year in the world literature, it can be said with confidence that none of those examined in this review (...) provided enough information to allow decision-makers to judge the distributional consequences of alternative resource allocations."* [17] This overall picture certainly did not change in the last decade. Yet this implies that the large majority of these studies do not help very much in making the basic challenge transparent. Remember that this basic challenge was how to cope in the most equitable way with the increase in health care costs.

A second problem is that the cost-effectiveness method neglects the basic trade-offs between different values. Simply computing cost-effectiveness ratios is not sufficient to take decisions, if one does not explicitly introduce a critical value demarcating 'effective' and 'ineffective' interventions [18]. And where could this threshold value come from if one refuses to introduce willingness to pay?

If the new interventions have to be included within a fixed budget, a transparent technique should indicate what other interventions have to be removed. When NICE in the UK decided that the cancer drug Herceptin was cost-effective, practitioners in the field were immediately confronted with the question of how to finance the expensive drug. In a remarkable paper, Barrett et al. (affiliated to the Norfolk and Norwich University Hospital NHS Trust), give an overview of different possibilities within their specific hospital environment and then formulate the problem in a very direct way: *"No patient is anonymous, especially not to the attending doctor who also has the ultimate rationing responsibility in the current system. (...) We, not NICE, have to choose which other treatments will not be provided and which of our patients will not be treated. Nobody has suggested what treatments we cut in favor of*

---

17 Sassi, F., L. Archard, and J. Legrand. 2001. *Equity and the economic evaluation of health care.* Health Technology Assessment 5, no. 3.
18 There is a huge body of academic literature on this topic. See Gafni, A. and S. Birch. 2006. *Incremental cost-effectiveness ratios (ICERs): The silence of the lambda.* Social Science and Medicine 62: 2091-2100, for a position which is close to mine.

Herceptin – not the media, medical advocates of the drug, the courts who upheld patient appeals, or NICE. It would be especially interesting to know what the Secretary of State for Health would like us to cut."[19]

If on the other hand, the health budget is to be extended, then the whole discussion about economic evaluation should be embedded in the debate on the desirable growth norm. Different medical innovations and their added value should be compared with each other and with intervention methods that are already reimbursed. And what is more: they should be compared with other allocations of the government budget for issues such as education, housing and poverty relief. This requires that the idea of willingness to pay is integrated in the analysis, that the relevant trade-offs are made explicit and that distributional considerations are taken seriously.

Cost-effectiveness studies are useful in structuring the debate on new drugs and new medical interventions. But more is needed to start tackling the real challenge. Courage is needed to formulate explicitly and in a transparent way the relevant taboo trade-off between material welfare and health and to draw attention to the much-needed solidarity.

Conclusion

I have argued that there is a considerable willingness to pay for health care. The real challenge is how to organize the system in such a way that not only the rich can benefit from the future technological progress. This is a real challenge in the light of evidence suggesting that there is a decline of solidarity in our Western societies. I have suggested that we should try (1) to focus on the insurance aspect, and (2) make the process of priority setting as transparent as possible. The discussion about the growth norm should be broadened to avoid number fetishism. Methods of economic evaluation should be refined so as to include explicitly distributional considerations and trade-offs between health and material welfare.

---

19 Barrett, A., T. Roques, M. Small, and R. Smith. 2006. *How much will Herceptin really cost?* British Medical Journal 333 (25 November): 1118-1120.

All this should not be seen as a plea for letting the expenses increase in an unlimited way. Nor did I intend to downplay the importance of increasing the efficiency of the system. I wanted to draw attention to the broader social context. How to translate the overall willingness to pay in an equitable system of public health insurance? What should be the future role of private supplemental insurance? Who should play an active role in the development of 'managed care'? What is the role of sickness funds and other players in the regulation of the expenses? There is much room for reflection and for improvement – but the need for transparency is crucial. Everybody active in the sector should formulate the trade-offs as explicitly as possible. Only in this way can we confront the population with the choices that we need to make, instead of stashing them away in elaborate technical reports that only a few experts can really understand.

BIO

Erik Schokkaert (KULeuven) is a prominent Belgian economist and Head of the Research Center of Public Economics and in charge of the Centre for Economics and Ethics. Schokkaert is Full Professor at the Catholic University of Leuven since 1995. He teaches Economics, Health Economics, Introduction to Public Choice Theory, and Welfare Economics. In his research activities, he focuses on distributive justice, health insurance, optimal taxation, social security and altruism.

ERIK SCHOKKAERT

[113]

NOTES FROM A TRAVELER ON THE ROAD TO CELL THERAPY IN DIABETES

DANIEL PIPELEERS

For many years, my research has followed a road map that takes laboratory studies on the biology of insulin-producing beta cells as a guide to head for a clear destination, the correction of beta cell failure in diabetes. It was realized from the start that the journey would be long and risky, with obstacles difficult to cross by lonely rangers and with an endpoint that may take more than a lifetime. As a long-time traveler on this road, my insights have gained from scientific progress in our projects and in those of others; they also remained – I believe – conscious of the constant need to challenge and to extend current knowledge. My views have been influenced by the climate in which our work was undertaken, and thus by its habitat and micro cosmos, by its region and by its weather makers. They of course also reflect the color of my glasses. One day I may write down my account of this journey, not because the world is waiting to read it but to settle and order my memories for myself, just as I like to arrange my photographs of family and social life, of travel and nature. This day may never come. In the meantime, there are sometimes circumstances when I voice fragments; for this CROSSTALKS book, the organizers asked me to produce a brief text, which should thus be considered as an excerpt from my travel notes.

Need for beta cell therapy in diabetes

Diabetes is a frequent chronic disease that reduces quality of life and increases the risk for life threatening complications despite current treatment. The disease can appear at all ages with the type 1 form being most prevalent under the age of 40 years, and type 2 becoming more frequent with increasing age.

Type 1 diabetes is caused by a massive loss of insulin-producing beta cells in the pancreas following a process that involves inflammatory and immune reactivities. Patients require life-long subcutaneous insulin therapy, either through multiple injections or through continuous supply by a pump. This life-saving treatment cannot however mimic the physiologic blood sugar control that is achieved by the beta cell population in non-diabetic individuals. Consequently, blood glucose levels can widely vary, down to levels that acutely cause unconsciousness and sometimes coma, or up to hyperglycemic episodes that initiate tissue lesions that can eventually result in serious chronic complications such as blindness, kidney failure, non-healing foot lesions. While patients differ in the occurrence and severity of their symptoms, they all face the daily burden of their disease. Against this background, a cure for type 1 diabetes is long since considered a priority in medicine. There is also general agreement that this requires restoration of the lost beta cells.

Type 2 diabetic patients still have a considerable number of beta cells in the pancreas but their mass and functional state have become inadequate to meet the metabolic needs of the body; in many cases this shortage is caused by increased overall needs as occurs in obesity. Weight reduction, changes in life style and hypoglycemic oral drugs can improve metabolic control in many patients. When these measures are, or become insufficient, methods that increase the beta cell mass also become a treatment of choice for a number of type 2 patients.

Both forms of diabetes could thus benefit from therapies that restore a normal functional beta cell mass, which makes the disease a top target for cell therapy and regenerative medicine. If such therapies would become available, and found efficient and safe, they would face a huge demand in view of the size of the target group, which, in Western countries, can be conservatively estimated at 0,1 to 1 percent of the entire population.

The need to restore a normal beta cell mass in diabetes, in the first place the type 1 form, has stimulated over the years many research teams towards two distinct strategies. The first, and the longest in development, intends to replace the deficient tissue by transplanting an insulin-producing graft. The second aims for a regeneration of beta cells in the pancreas. The term beta cell therapy is used for interventions that aim at restoring a functional beta cell mass; it thus covers beta cell transplantation and beta cell regeneration. The two therapeutic approaches are not mutually exclusive as they may apply to different phases in the disease, and may also be considered in combination. They furthermore share criteria that should be met in order to achieve long-term correction of the diabetic state. In general terms, both strategies should restore and preserve a beta cell mass that mimics the functions of a normal beta cell population. Knowledge on beta cells thus represents a key element, which explains our long-term interest and activities in this field.

Need for a collaborative network

The Diabetes Research Center (DRC) of the Vrije Universiteit Brussel (VUB) has a long-term interest in the biology of insulin-producing cells. Our team has been the first to purify beta cells and to investigate their properties in isolated preparations. This knowledge has helped us to assess and understand the *in vivo* function of these cells, and to design strategies for beta cell therapy in diabetes. The intention to translate our findings to the clinics, raised the need for bringing in complementary expertise and collaborations. Over the past 15 years, VUB/DRC took initiatives to build an international multidisciplinary network of basic and clinical research teams that aims towards beta cell therapy in diabetic patients.

A long-term program was outlined in which clinical trials are guided by a biologic platform, and supported by reference centers and bio-industrial partners (see figure on page 118). It has been recognized and supported as a Center of Excellence by the European Union (since 1990) and the US-based Juvenile Diabetes Research Foundation (JDRF) since 1995. It forms since 2002 the profile of the JDRF Center for Beta Cell Therapy in Diabetes (www.betacelltherapy.org) with its central unit on the medical campus of the VUB and with the DRC as a major partner.

Composition of the JDRF Center for Beta Cell Therapy in Diabetes

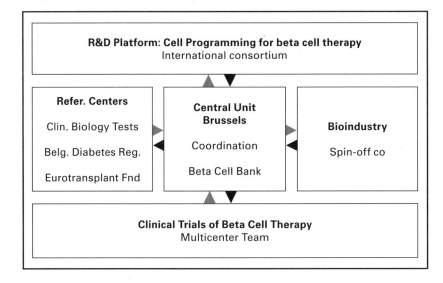

Two long-term trial lines have been initiated. A multicenter team conducts them, with a major participation by the Belgian Diabetes Registry and its affiliated university and non-university hospitals. *One trial line* has shown that it is possible to protect beta cells against losses when a short-term antibody treatment is given at the time of clinical diagnosis; further work now examines how this effect can be amplified and extended to earlier phases. *In the other trial line* we have demonstrated that diabetes can be corrected by beta cell transplantation in patients with early stage complications. The human beta cell grafts are prepared by our central Beta Cell Bank, which receives donor pancreases through the intermediacy of the Eurotransplant Foundation. The size of these cell grafts is however too small to induce a long-term metabolic normalization; furthermore the number of available grafts is largely insufficient given the number of potential recipients. The R&D platform therefore seeks methods for mass generation of beta cell grafts in the laboratory. Such technology will help us to set up larger scale trials that identify the conditions in which beta cell grafts achieve a long-term efficacy and safety. It will subsequently allow implementing successful protocols as a cure in numerous patients.

Several potential sources for mass production of beta cells are currently under investigation. A procedure has been developed for isolating large numbers of perinatal pig beta cells; this type of graft has been successfully tested in mice and is now being studied in large animals. Other approaches focus on reprogramming human cells to insulin-producing cells. Along this line, one of our partners has derived insulin-producing cells from human embryonic stem cells and these are now being studied in our pre-clinical models. Two other collaborating teams are developing techniques for producing beta cells from other cells in the body, which would offer the advantage that cell grafts can be prepared from the patient's own tissue, thus avoiding the need for immune suppressive treatment. These different approaches to mass production of beta cells are undertaken in parallel. They involve close collaborations between university teams and small bio-industry enterprises.

Need for a road map

Subjectively, I believe our work has led to significant progress towards the clinic, as well as to promising findings and new developments in the laboratory. These achievements have been the result of a road map that took cell biology as a guide. Taking steps on this road map, and reaching milestones also indicated several other needs. It also generated side observations and reflections, some of which are briefly stated here. They may apply to other areas as well and then deserve broader discussion.

In our case, translation from the laboratory to the clinic is a demanding, long-term and costly undertaking. This is only justifiable if the intended clinical benefit is clearly defined, sufficiently important and realistic. The goal of restoring a functional beta cell mass in type 1 diabetes meets in our opinion this requirement. The issue is then to find the resources for this long-term effort in collaborating disciplines. This is not evident in a country that has paid little attention, and money, to translational biomedical research, and that does not provide financial resources for academically driven clinical trials. In addition, Belgian universities and hospitals do not have a tradition of recognizing each other's strengths and collaborating along this criterion. That our program has been able to progress is the result of resources that we succeeded in finding at three different levels.

The first level is that of the DRC/VUB, where our launch has benefitted from a climate favoring the development of novel initiatives on the new medical campus. Excellent scientific and technical collaborators have been recruited. They have built, over the years, a unique expertise that is essential to the program. A continuous challenge is however finding the resources to maintain this critical mass. Continuity of core expertise is essential and yet not easy to realize, a problem that is certainly also encountered in other Belgian teams but that is felt more strongly at the smaller universities in view of their unfavorable funding by the government. Since the start of the DRC/VUB, its projects have been supported by the Flemish Science Foundation (FWO), which does not, unfortunately, manage budget sizes that meet the basic needs of laboratories. Recent help on this issue has come from a Methusalem grant given by the Flemish government.

A second level is the commitment of most Belgian departments of clinical diabetology and transplantation to collaborate with our program; in fact only one university (UCL) refused and wanted to go on alone. The Belgian Diabetes Registry (www.bdronline.be) plays an important role in creating and maintaining a collaborative clinical network of university and non-university clinical departments in Belgium. This second level of interaction is strongly driven by diabetologists as well as by patients and their family members. It represents a unique instrument for clinical studies and for diffusing information to (para)medics, patients and society. Unfortunately its funding is fragile and problematic. Also striking is the discrepancy between the interests and active participation of French speaking patients, nurses and doctors and the disinterest of their public health department.

The third level is the international network funded by the European Commission and by the JDRF. These five-year grants support international collaborations with excellent teams. They made our translational program and the clinical trials possible. They also made us realize that the cost of scientists and technicians in Belgium is 15 to 70 percent higher than in other countries so that our foreign collaborators can hire more personnel with an equivalent budget. A considerable fraction of our foreign grants goes to taxes and overheads. When we calculated this fraction for the year 2006 we found out that it amounted to 50 percent of the grants that we received from Belgian sources, in other words we were paying back half of our local grants through taxes and overheads on our international grants. Such a situation is opposite to that in other countries where foreign grants are often matched with local funding, brought in by research institutions and foundations. In the case of programs with medical relevance, it would also make sense that the public health system invests in particular when trials are planned for preventing and curing diseases that are responsible for major expenses. Diabetes presently absorbs 15 percent of the health budget; purely for economic reasons, it would seem logical that a fraction of this cost is invested in efforts that should ultimately lead to a reduction of this cost.

Role of stakeholders

Traveling the road has raised questions on the role and place of stakeholders. Patients with diabetes are undoubtedly the most important stakeholders, as are their family members, and individuals at risk of developing the disease. We therefore regularly inform them on the perspectives, the progress and the obstacles of our program. Many of them participate in studies or hope to be included in promising intervention trials. All of them want investigators to move fast on the road and expect policy makers to create the appropriate conditions. As the second category of stakeholders, investigators should design the clinically most relevant road map, request the resources and report on their use; their activities are continuously monitored and controlled from different sides and angles. Such review does not however exist for the policy makers, who not only have responsibilities towards patients but also towards the public health budget. In fact, their stakeholder role and place should be clearer defined and carried out. The fourth category of stakeholders is composed of the hospitals and the pharma-industry, both of which deliver services and products for diabetes treatment today and will also do so tomorrow. Their agenda is heavily focused on care and products of today, and on budgetary targets. Their responsibilities in finding ways for prevention and for durable treatment cannot of course be denied; mechanisms that ensure their role may thus have to be installed or strengthened. A road map is a useful framework in which to locate all stakeholders. Travel along this road is thus in no way a solitary walk.

BIO

Daniel Pipeleers obtained his MD and PhD degrees at the Vrije Universiteit Brussel (VUB). From 1971 to 1980, he was Fellow and Senior Investigator of the Belgian Scientific Research Fund (FWO) in the departments of Prof Willy Malaisse (Université Libre de Bruxelles), of Prof David Kipnis and Prof Paul Lacy (Washington University, St Louis, USA) and of Prof Willy Gepts (Queen Elisabeth Foundation, Brussels). Since 1980, he is Professor in Pathologic Physiology and Biochemistry, and Director of the Diabetes Research Center at the VUB. His team examines how the biology of insulin-producing beta cells can be used for developing and implementing beta cell therapy in diabetes. With this objective he has launched an international collaborative program that is supported since 1991 by grants from the European Union, the Juvenile Diabetes Research Foundation and the FWO. From 2002, these activities are organized by the JDRF Center for Beta Cell Therapy in Diabetes of which Daniel Pipeleers is director. He is also founder of the "Belgian Diabetes Registry", the "Beta Cell Bank", the VUB spin-off companies "Beta Cell nv" and "OPUS nv". Pipeleers is a member of the Belgian Royal Academy of Medicine, Doctor Honoris Causa at the University of Uppsala, Sweden, and Methusalem investigator of the Flanders Government. His work has received several awards, among which the European Minkowski Award, the Quinquennial Prizes of the Belgian government and of FWO Flanders, and the International Bial Merit Award in Biomedical Sciences (Portugal).

[123]

SHARING RESPONSIBILITY IN THE DISCOVERY, DEVELOPMENT AND DELIVERY OF MEDICINES

A ROUNDTABLE SESSION ABOUT THE ACTUAL CHALLENGES AND POTENTIAL ADDED VALUE OF AN INDUSTRY IN TRANSITION

MARLEEN WYNANTS

A new paradigm in health care is based on the thesis that health is a shared responsibility by all stakeholders: policy makers, employers, cure and care practitioners, pharmaceutical industry, patients and their organizations, health insurance companies and sickness funds, as we call them in Belgium. Health care quality and affordability can only be realized through a thorough reform of the actual system that has reached its limits in every possible direction. Trying to find ways to put enough money into the current system is impossible since the system is designed to expand and to absorb the amount of money available and will continue to be under-funded. Solidarity of society as a whole and taxpayers and patients individually, linked to a true responsibility of every stakeholder, are the two main pillars of an optimal and efficient health care system. And whereas the pharmaceutical industry remains a major player in health care, a global solution will not stem from continuing to cut budgets on the development and delivery of medicines only. Rather, a possible solution will emerge from a new and more collaborative development process, embedded in a health care system where risks, responsibility, savings and gains are proportionally divided among all stakeholders.

A book on the future of medication and its role in patient-oriented health care would be incomplete without taking the perspectives of the pharmaceutical industry into account. Hence CROSSTALKS organized a round table discussion with the major players of the Belgian pharmaceutical industry. The participants were **Julien Brabants** (Executive Director Public Affairs, GlaxoSmithKline), **Christian De La Porte** (Medical & Public Affairs Director, Janssen Pharmaceutica), **Piet Schutyser** (Vice President and Administrator – NV AstraZeneca SA, CEO AstraZeneca Foundation), **Luc Vermeesch** (Vice President Europe, UCB) and **Kris Westelinck** (Managing Director Pfizer Belgium/Luxembourg).

The limits of the health care system

Luc Vermeesch: The perception at this moment is that whether we talk about the financing of the whole system or the development model of new medicines, the limits of the existing health care system seem to have been reached. The budget is increased on a yearly basis by 2 or 3%, despite claims that it should at least keep track of the increase in GDP, although nobody believes that this can be maintained. On the other hand, no other alternative yet exists to keep costs affordable. So we have arrived at the limit of a system and the question is whether this is a real limit or whether there is another way of looking at the situation. The same goes for medicines: a lot of medicines enter the market whether from private, public or collaborative initiatives but the question is: When is the incremental advantage of a new medicine enough for it to be accepted? The expectations of improvement between something new and something that already exists have dramatically increased. Hence the development model for research aiming at new products, whether these are breakthrough products or better versions, is also at stake since today we are supposed to mainly work on breakthroughs. And we all know breakthroughs are exceptional. So how can we maintain an economically viable R&D model with continuity of funding while not only relying on breakthrough medicines? How do we deal with that? We won't solve the problem by simply giving the system a good shake up and increasing the budget by 2%.

Kris Westelinck: Indeed, if you look at the health care system today, we see a system that worked fine until halfway in the 20th century. But needs have changed. And patchwork will no longer solve the emerging problems. We are evolving much more from cure to care, and that implies gigantic amounts of money. Hence a new system has to be developed that is in tune with the needs of the patient. When you consider the role of the pharmaceutical industry in this, we know that breakthroughs are exceptional, but if you take the incremental steps out of the system, if these small steps forward can no longer be paid for, then you will miss out on the major steps in the long-term. And that's the dilemma we are facing today.

Christian De La Porte: For me, the first problem is that almost any debate on this issue is fed by the idea that the budget can only grow by a certain percentage. One considers the whole health care system only from a cost perspective, not from a benefit perspective or an income perspective. Secondly, the present business model is coming to an end for various reasons. One is the graying factor, meaning that we evolve from an acute care model to a chronic care model, which has a different cost factor but also a different benefit factor. Then there are new products that will become more expensive if you throw out the incremental products and narrow down the market. One keeps relying on fewer means and there is no balance between the different points of view. I'm still convinced that our task is bringing new products to the market, innovative products, but there is a panoply between incremental products and breakthroughs, a circle assuring the income of a company, and if you hamper that, you cut the current. And that is what is happening at this moment.

Piet Schutyser: Whether the system has reached its limits is not the most essential for me. The key question remains whether the limits of output have been reached. I'm convinced that the limit of how health care system resources are used has not been reached. This is the first thing to investigate and if the answer is no, then there is no problem. If yes, then there is a social and political responsibility to define health objectives and the budgetary needs associated with them. We need a social debate here about who pays for what and why – and of course, based on solidarity at the level of *payability* and access. Only when all stakeholders, the pharmaceutical industry included, ask themselves these questions, only then will it become clear what the framework is within which we can realize our corporate ambitions. Our role is to offer innovative clinical solutions to existing medical needs with respect to social choices and challenges within the collectively established health care framework. Our perspective on and consequently our initiatives in directing the system and our impact on it financially will reflect the degree of real respectful partnership and a shared sense of responsibility. The objective is to aim at a social balance between our corporate ambitions and our clinical promises for an optimal quality of life for every individual.

Julien Brabants: I want to add another layer to this discussion: I'm convinced that a society should only pay for what it thinks has a value. Today it is very hard to come up with totally new medicines. That is true. When we state that the model doesn't work anymore, I have the feeling that we already said that ten years ago and in the meantime, it still stands. Yes, with loose ends but what is the alternative? More importantly, in my view, is that society is taking a more and more strategic view of health care management, and is focusing increasingly on value for money. That's why payers may increasingly make reimbursement conditional on performance. At GSK we are experimenting – and maybe that sounds naïve to some – in involving society quicker in the development process. Already during phase I, we set up a forum with the main insurance companies in Europe, that is the sickness funds and people like Noël Renaudin (Président du Comité Economique du Médicament, France) and Michael Rawlins (Chairman of the National Institute for Health and Clinical Excellence – NICE, UK) to give them an overview of what is in our R&D pipeline. And I'm not talking on an abstract level; for example, during the last meeting we presented a series of compounds to them and about some they clearly said: *"That's a great theoretical concept, but we are never going to pay for that!"* I think we have to dare to ask society what it needs from us. I don't say we have a solution yet; we are just exploring some possibilities. Basically I'm also in favor of funding for incremental innovation, but how far can we go there? I think medicines can be fine-tuned and improved step-by-step and something that is of no value today can suddenly mean a lot in five years time, so we should be flexible about that. In this sense, values should be re-estimated, because nothing is static.

Schutyser: For me it is essential to merit the success of our medicines through real added value. It is our duty that when we launch a medicine on the market we merit the use and the price we ask. On the other hand, the system should stimulate and re-compensate our sector to investigate the unused potential clinical value in existing medicines, to accord patents on new indications and adapt the price in function of the created added value.

De La Porte: Be careful about asking society what it wants because once you know that, you will start looking in a specific direction and you end up in a narrow tunnel. Whereas experience teaches us that you might discover valuable sidetracks before actually stumbling upon something really innovative 10 years later. Does it also mean that every time you find something, you have to go back to society to see whether society wants it or not? Where I do agree with Julien is that within society you can indicate where you want to go, backed up by public debate where patients do have a say, which is not always the case in Belgium.

Vermeesch: We have 'society' and insurance; both are not necessarily the same. As industry is part of the health care system, you cannot make an abstraction of the client, whether it is a patient, a doctor or an insurance company. Maybe that happened more in the past, but today we are confronted with new elements and I think Julien is right in saying that we don't yet know how things function. I remember a recent discussion when we reflected on research and we concluded that about one third of research is pure private research funded by the pharmaceutical industry, and this is the fundamental 'driven' research covering the sidetracks Christian is talking about. Here you will find things that were not on the agenda or part of the deliverables but they remain of high value. Another third of research should in one way or another take into consideration what society wants or needs, whether society is the sickness funds, state, or patients or both. In the context of the European Medicines Agency (EMEA), patients do have a voice, and if not systematically, then certainly occasionally in evaluations or preliminary discussions. There are real implications here since you can no longer make an abstraction of your context, whether you are involved as a public authority, pharmaceutical company or sickness fund. And yes, opening up your portfolio to parties that were not previously engaged in the discussion is a step in the right direction. Whether this is going to deliver solutions is not important, it's the process that opens up. The last third of research will have to be embedded in the North/South issue and thus part of a longer term and global vision. For me these are the three axes on which the actual R&D model will have to focus and direct its efforts and organization.

Schutyser: I want to again stress the unconditional priority to investigate the correct use of resources. On the other hand, political debate should be more transparent and a real reflection of the collective need and concerns. Finally the model of budget 'silos' should disappear. When one therapeutic approach offers real savings, this should then impact the budget of the substituted act.

De La Porte: We have to be careful not to confound our clients since the client is the patient and NOT the insurer. There are thousands of diseases where solutions exist and about five times as many where no solutions have yet been found. And if I learned anything from being a doctor, it's that when you wake up in the morning, you will face about 100 problems during the day that follows. If you wake up ill, you only have one problem. And that can be a problem with an existing solution or a not yet existing solution, but that problem still has to be solved, whatever your insurance company thinks about it.

Brabants: I agree. What I wanted to say is that we are all working in our own narrow corridor right now and are jubilant when we find a nice receptor. So we start developing and in the end we come up with something and we confront society with it – doctors, patients… – and we have to start convincing them of something we believe is intrinsically valuable. So why not do the opposite? Share what you found to start with, discuss it together and see whether it can be of any value? One of the main frustrations of pharmaceutical companies, is being constantly confronted by other parties – whoever they may be – who are unwilling to pay for the things we develop. Neither do they consider them as a need in the first place.

Westelinck: *"Whoever they may be"* – I don't like that. If a sickness fund has to valorize that, it's becomes a budgetary issue and we remain stuck in the trap we're in today. You also stated that about 10 years ago we already said the system was no longer tenable and that's possible. But today we must admit that it is no longer tenable and maybe some patchwork will help for the next two to five years but it will crash if we don't adapt the system to the needs of a changing society. And I would rather start doing this today than waiting until that is the only option left.

Reforming the system

De La Porte: As an industry, you expect from society a specific vision on health care that looks towards the future with deliverables and milestones. For example, when society says: we need to cut the number of amputations due to diabetes by half within five years – well that is a specific aim. You plan retroactively up until prevention, but at least you have a goal and you know you'll get there. And everybody, industry, mutualities, cure and care providers, all stakeholders know exactly the role they will have to play. But society has to develop that goal and I think that is lacking in Belgium and in other countries. Some countries are capable of doing that, sometimes with painful consequences, like if you develop gangrene in your legs in Australia but you refuse to stop smoking, you can forget your amputation. You live within a system and you know the rules and conditions in order to obtain health care. It may seem hard but in Belgium, certain patients are also denied their medicines but it is less transparent or visible. And that is totally unethical since everybody in the system with decisive rights, escapes responsibility.

Schutyser: I would like to call for stopping the divided approach in the decision-making process in the health care system. Each partner – sector – has an agenda and acts according to its own objectives. A mixture of corporate, societal and medical ambitions drives the daily decision-making process. Of course all nuances must be present in the process, but the question is when and at what level? The basic platform should be clinical and asking the essential questions: *"What do we want to achieve, what are the priorities, what are the real tools today, where needs are unmet, what do we need to solve them?"*

Westelinck: For me, a major part of the reform needed is the embedding of health care in society, since today it is nonexistent. Seen conceptually, the Belgian Health Care Knowledge Center could have been a possible actor in this, but it evolved in a different direction. So what is the alternative decision-making process? You cannot possibly take the statement *"We have to keep this year's budget balanced!"* seriously any longer. That is far from a moral approach to health care! The pharmaceutical industry represents 13% of the total cost in health care. Even if we delivered our products half price, the government would still not have enough money since there are so many inefficiencies in the rest of the system that has not adapted to the changed needs of society. We have the graying of society, more technological drugs… Merely raising the quality is not a parameter. The change from cure to care is costing a lot of money and will cost even more in the future. Just look at staff costs in the health care system; these will keep rising and are inevitable. But if for all these reasons no debate is held, the industry can half it prices and we'll still be the first to be targeted. Industry's share is 13%, medicines are 18% – part of which is distribution and VAT, and not even directly related to us. But expecting that 80 to 90% of the savings government wants to realize has to come from the pharmaceutical industry is unrealistic.

De La Porte: It's true that the government wants to save €400 million and that €280 million will have to come from medicines. This is way out of proportion. Consider psychiatry: when you treat a patient, 5% of the total cost in Belgium is medication; 95% is other things. So as long as no one is prepared to do something about the other 95%, the system will never become better balanced or more efficient. And a major obstacle is that the expected advantages of system reform are more than one legislation away... Moreover, you can never transform the system as long as it remains based on a monopoly. And the health care system in Belgium is stuck in a kind of monopoly in the sense that you as employer, worker, patient, whatever, have the obligation to insure yourself and to pay for it without having any say in what is in the contract. You don't have the choice to directly pay the person or organization you want to pay. Every item or change in the insurance policy is removed from the payer or from the person the policy is destined for. This is unique at global level.

Schutyser: Perhaps, but my point is that we have to merit this 13 or 15% share of the total health care budget. Full stop. We have to concentrate on that.

Vermeesch: The system will never survive by savings on medicines alone. I think there are two alternatives. First, there is pressure, and not only in Belgium, on the innovation cycle and this will have to be reconsidered by the industry itself. Second, what about the macro context of health care and financing? And that goes far beyond medicines only. So if we want to come up with solutions on a macro-economic level and meet national health care objectives – cure and care objectives – then the first initiatives will have to be conceived at that macro level. With regard to continuity for the pharmaceutical industry itself, nobody will offer us a solution so we are going to have to scratch our heads and start thinking.

Westelinck: I agree. We have to look at our own system and try to improve our own efficiency but expecting that we will save the whole system in doing so is an illusion. A total vision or common future perspective by all stakeholders is lacking. We have to pick up the challenge and sit together to continue an open dialogue like the one CROSSTALKS set up and keep developing transparency between all stakeholders. We have a major role to play there. But the other stakeholders must recognize the industry as a partner – like Guy Peeters[1] stated – mistrust should disappear. We have been ignored for far too long although we are a very important player.

---

1   See article by Guy Peeters, page 167.

Vermeesch: Setting objectives is one solution. However, this is a learning process and not something you can just come up with. Why have some countries effectively done so? Because of economic reasons, otherwise they would never have the power of enforcement.

De La Porte: Still, the fact that it will be a painful process, is not a good reason NOT to do something and so far there has never been an inquiry in Belgium about what the patients, the citizens are willing to pay for. Everything is supposed to be free and if you confront people with brutal choices, they don't know what to say. And even if it has to be free for some people that does not mean it should be free for everybody.

Schutyser: Credibility is the basis for everything. Criticism mostly targets how we do things, what we are standing for. So let's start by promoting the right medicine for the right patient at the right moment. This implies responsibility at all levels of HCPs, patients and ourselves included. Furthermore, how do we want to see the system progress? Full solidarity and shared responsibilities or free choice for the individual? To develop a mixed system with a minimum health care service packet for social cases could be a nice challenge. But how to guarantee the best care for all?

An industry in transition

Brabants: We shouldn't over-generate – questioning people does happen although it may happen a little too much from a prejudiced perspective. But look at the recent analysis of the Christian Mutualities[2] – some people don't have the means to go on holiday or buy a second car or pay for health – for those people medicines and care should be free, there's no way around it. My point is: What is the added value of the medicines we produce? We should look very critically at ourselves. At one time, you could avoid stomach operations through H_2 blockers. Initially these blockers were pretty expensive. They had to be because R&D investments had to be compensated. But today R&D costs are covered and only production costs remain (a tiny fraction of the initial costs). So we should reconsider and that is exactly the debate we don't yet dare to enter. Also incremental innovation is very flexible in time.

---

2   *Sociale ongelijkheden op het vlak van gezondheid: vaststellingen op basis van de gegevens van de ziekenfondsen* – Hervé Avalosse, Olivier Gillis, Koen Cornelis, Raf Mertens – Afdeling Onderzoek en Ontwikkeling, CM, Juli 2008.

De La Porte: At the beginning of the discussion we stated that cost cutting is the one and only perspective from which things are considered. What you try to save on one level, is not being injected back into the system – so the saving in itself doesn't mean anything since it doesn't add anything to health care, it's just a means to keep the budget under control. Instead we should reconsider incremental innovation versus breakthroughs, reconsider investments and reinvest elsewhere, making sure medicines come to the market according to the needs of the moment. But the mechanism we have now is only cost reducing and cost cutting. And it's just like David Kudla[3] said: *"You can't cost-cut your way to prosperity."* It just doesn't work.

Schutyser: The industry must make clear choices for the future. Either it integrates itself in accepted societal models of solidarity and equal access to health care for all and fits in the defined model (preferably designed with its input), or it goes its own way of purely industrial ambitions, fulfilling patient needs without any concern for the equal basic health care rights of the individual. Ergo, it chooses public funding and contributes by every means available to make this approach work. If not, it should not call for assistance from social security funds if the model fails.

Vermeesch: Another issue is the necessary transition we as industry are facing and how we tackle that. Today many studies in late phase III are being stopped because either they don't come up with the necessary results or are not different or innovative enough versus what already exists. So you remove the research. But how can we as an industry make the transition towards a model that offers more guarantees on the level of 'improved applications'? Fundamental research is the less costly factor since we have different formats of collaboration, with public as well as private organizations. The question is: What do you do at the level of applications, in our case called clinical development? It's exactly at that level we need an interaction with our clients, with policy makers to be able to realize added values.

Brabants: I couldn't agree more. The industry should enter into dialogue much quicker with its partners to decide whether to continue developing or not. The main question remains: Are we bringing enough innovative products to the market? Since this is not the case, we're facing hard times. Let

---

3 David Kudla is CEO and Chief Investment Strategist of Mainstay Capital Management, LLC.

me go back to the H2 blockers: when the first H2 blocker reached the market, there was almost no alternative to stomach surgery for ulcers. The price was high, let's say 100, but society profited from it, so it was justified. Then other H2 blockers were developed and were also better, so the price rose to 110, 120, 150... But what if the price of the original H2 blocker falls back to 10? Should the rest remain at 120 and for how long, etc? That's an exercise we must make in the near future, because *mutatis mutandis* the same reasoning goes for other classes of medicines such as PPIs, statins...

De La Porte: Yet it goes beyond only medicines. I agree with Julien in the sense that when a certain product becomes post-patent, it experiences a devaluation in price. Take the H2 blockers that indeed now cost about 10% of their original price. But if you were to take the H2's off the market and start operating on people again, they would not be operated on at 10% of the original price. That is also part of the discussion we have to enter. We are talking about intrinsic value but if the rest of the market in health care does not follow the same principles, than your exchange value is at loss. So if you save money on medicines, the money has to be reinvested elsewhere in the system since the system itself does not devaluate. If you need a stomach resection, you can't possibly get that for the same price as 20 years ago and certainly not at 10% of that price. That's reality.

A system based on shared responsibility

Vermeesch: One question remains central to the whole issue: How can you keep up with a system that claims it is based on solidarity and where you can add things through private insurance or a supplementary one without bringing the solidarity part into danger? As long as we can't answer that question through tough debate, we'll still be telling the same story 20 years from now. You can't possibly exclude solidarity. I speak for the European platform in general. If you take solidarity away, you'll run against a wall. But how to integrate a private insurance aspect while trying to avoid a health care system at two speeds or without creating discrepancies that no one can solve? It's not about the willingness to pay according to me; it's about transforming the ability to pay into a system in such a way that it does not endanger solidarity in a fundamental way.

Westelinck: That's right. One player cannot change the system; policy makers included. You really have to bring all players together and organize a major social debate.

De La Porte: But you will only be able to bring them together and oblige them to change once you take away their power and impose one condition: you can't leave the room without finding a solution.

Brabants: Society itself will ask for a solution and direct it. You have to ask the people whether they are satisfied with the system, what they need more of or what should be improved. You will never be able to change something if the people are happy with what exists.

De La Porte: Don't forget that people are easily satisfied with something that is seemingly offered for free. One way or another we will always want to offer a basic basket of cure and care and you provide one that is larger for those who cannot afford more and a smaller one for those who can and who will buy additional insurance not in the main basket. Why not?

Brabants: As the devil's advocate I say: That's already what's happening today!

De La Porte: Well, I'm not so convinced about that. Don't forget that insurance today tends to take the position of "*We cannot pay this for everybody, so NOBODY will have access to it.*" That's completely immoral.

Schutyser: Society has to make a choice. Shared responsibility must be inspired by trust and willingness for partnership among all partners, patients included, sharing the common objectives in a transparent and non-opportunistic way. At the same time, our industry, if we want to be a part of that responsible partnership, must earn its trust. Ergo, we have to prove day by day our added value and convince society of our credible partnership, sharing common challenges and responsibilities.

Brabants: You can always bring a medicine on to the market without it being paid back. We all have a lot of stuff in our portfolio that is not being paid back, but yet it is available.

De La Porte: But when you consider health care in general, more than just medicines, then you realize that certain medical operations cannot happen because the material is not available. That's true of health care insurance too. Some procedures are automatically included in your policy and you pay for them, but you have no control over the policy itself. Some countries do have solutions that are better but then you unavoidably end up with cherry picking and that's something we all want to avoid. In that sense you have to face a tabula rasa, get together with the right people and start reflecting upon a new system. Keep adjusting some buttons to regulate the machine, a little less fuel, less money, a little more of this, decreasing the doses, stating that instead of three vaccinations, two will do… either we are talking science or we're not! Or the policy makers are wrong, or we are wrong or another stakeholder is wrong – only one party can be right here. And the discussion has to be free from any respective conflicts of interest. So we do have to listen to the patients in the first place, but patient organizations should be organized efficiently and involved and engaged wherever possible. At the moment some patients – look at oncology – are overrepresented while others don't have a single representation in the whole system.

Westelinck: Most stakeholders are very familiar with the system, the insurance companies and sickness funds, the care deliverers, doctors, pharmacists, and the industry… When you compare countries, you see that the US puts 16% of its BNP into health and even that doesn't result in a better system at any level, so it's not just a matter of pouring money in. Erik Schokkaert[4] states that 10% is not the limit – although it is what we do now. You have to make the whole system efficient otherwise you just keep pouring money in and you don't improve anything.

Vermeesch: You can't deny the importance of this in the development of therapy and medicines. Patients in oncology are involved much sooner than in the classical model and as a consequence clinical development is adjusted in a positive way. Take 1,000 patients, do a placebo-controlled study… that's not the way it is done in oncology – the interaction is much more productive or at least more innovative than in the classical model.

---

4   See article by Erik Schokkaert, page 99.

De La Porte: I agree. I think that the universities play a crucial role here since at the level of oncology, universities think differently and their research is value-driven. And I think that is the right model to pursue since a placebo-based model does not work here. In that sense, the model is justifiable and better than the classical model we normally use.

Vermeesch: And it deals with general care from an early stage! If you consider pharmaceutical products, I think in Belgium and in other countries, little to nothing is being developed without the collaboration of a university at one stage or another. And this is increasing, so platforms do exist. The question that keeps coming back is whether on the level of valorization we could do more than research. The relevance of the collaborations could be enhanced and guided towards even better research, clinical research and development. Are these collaborations relevant for the payer or not? There is no complementary compensation for either or both, and that is something we should also think about.

De La Porte: Responsibilization remains pretty limited. Hospitals get extra funding but has there been any validation of what they offer society in a transparent way? A series of valuable and perfectly patentable developments at universities are gathering dust since nobody has the insight to take them one step further. Also a huge optimization imposes itself there. I go back to the beginning of this discussion: one should not only look at the costs but also at the benefits. University funding is being seen as a cost factor – money is brought in, but what comes out? This is not always tangible, which is fine, but next to research, papers etc., a lot of stuff is produced that could be taken one step further and could benefit society, create an income – but that circle is far from being closed.

Brabants: The only approach I can suggest is to involve people in what we do much faster and in parallel be more transparent and try to come up with collaborative solutions, but I don't believe in a tabula rasa. A society builds itself, learns, gets lost, corrects, adjusts... We'll see what emerges from our common efforts and go from there. We're better off doing this in the manner of Erasmus: *Festina Lente*. French revolutions are not exactly long-term movements. Robespierre was executed, wasn't he? I keep thinking that in the last 10 years we've simply not brought enough new products onto the market.

Vermeesch: And if you can't do that, you'd better stop what you are doing.

De La Porte: True, since that would indeed mean that everything is discovered and we have no added value to offer.

Westelinck: Which is not the case at all. We should remove inefficiencies in our own system, but the rest of the health care system should do the same.

Brabants: That's exactly what I think. We're going through hard times and I think we have to reconsider the way we do our research or development and adjust internal processes in tune with the academic world and the rest of society. And that will take some time – new things will emerge. That's what it's all about in the first place. We must reflect on what we can offer and also what we seek. We will deliver new medicines and vaccines that address unmet needs and have demonstrable value. On the other hand, we also seek increased focus on prevention, more strategic management of health care budgets, prioritization of improving health, elimination of inefficiencies, a focus on value not cost, faster patient access to new medicines and vaccines via new ways of negotiating, better dialogue between industry and public authorities, which should start prior to the marketing authorization, and putting patients at the heart of decision-making.

Schutyser: I fully agree, but why inform responsible partners on what we have to offer? Why not involve them in the decision-making process of R&D strategies? Shared responsibilities can't only be a mind-set at the moment when all items are fixed and without opportunities for change, but must be a 'common thread' running through the whole health care process driven by optimized consolidation as well as and/or innovation.

Westelinck: Well, the technology has changed a lot and the graying of society has yet to show it's real impact – it's only just begun. The model of cure to care is evolving NOW. So let's look at the processes: just by saying we need more money in care and taking it from another budget, is a short-term solution. The system should be transformed until it works better and more efficiently. And we as industry have to do our part but it doesn't make sense that we adapt our processes and enhance them since we only represent 15% of the whole system! Evidence based results are a condition for us, but what does evidence based mean for the other stakeholders?

De La Porte: Moreover, the impact of the other stakeholders – hospitals, insurers, pharmacists, cure and care practitioners – is much larger than ours. How many hospitals have a quality output on their own functioning that is comparable to other hospitals? Comparable quality information about health care just doesn't exist. As a patient, it's a real challenge to find the difference in the quality of cure and care services that the various insurers or institutions offer... and in line with what you have to pay. You have no possible clue where to go or what to negotiate in your insurance contract. Some stakeholders are not willing to give this information either, so transparency remains a big issue here. Rankings could be one of the instruments but it is even more important to work with good and relevant parameters whenever something is published.

Westelinck: Too little attention has been paid to quality. Everything is driven by budgets. And with regard to that budget, you can merely assess what the quality is that you associate with it. By enhancing quality and processes, you can probably use restricted resources in a better way. But then you need health objectives, as we stated before, and only then can you valorize them and make the results public. An independent Health Care Knowledge Center could play a major role in this but then its core mission would have to be defined as such. And not only with an eye to providing more information for the patients. A secondary advantage is the fact that those who don't score well, would have to work on improving. And those who score well would want to maintain their score. That could have a direct impact on cure and care, and that's what matters.

Schutyser: As I said earlier, shared responsibility is a continuous process starting from our R&D strategies and ending in how we 'promote' i.e. inform others about our medicines. The day we prove that the scientific level of the information is of high importance and that the content leads to added value for the patient, there will be no opposition. Today our communication objectives create a perception of 'pushing' prescriptions and do not always reflect willingness for shared responsibility. Validated clinical information on our medicines, leading to better health care service to the patient, should be the core of our information strategies. This would contribute to trust, evidence of responsible behavior and finally to a credible partnership with all stakeholders in the health care environment.

Brabants: There are still a lot of anomalies in the system, that's true. What always makes me frown is that we as the pharmaceutical industry cannot publish any information on medicines but if an individual does, that is legal. This is not correct or tenable since it's an inevitable step in the information process, as long as it is transparent.

Vermeesch: One of the toughest challenges in the whole information process and also in this initial CROSSTALKS project seems to be involving doctors and other health care practitioners. They play a major role and have the main responsibility in the whole process and system of health care. So let's not forget that most of the people confronted with illness will rely on their doctor. And regardless of how big and transparent the external part of information will become, the doctor will always remain THE point of reference in the system.

WITH THANKS TO

Julien Brabants
Executive Director Public Affairs, GlaxoSmithKline
Christian De La Porte
Medical & Public Affairs Director, Janssen Pharmaceutica
Piet Schutyser
Vice President and Administrator – NV AstraZeneca SA,
CEO AstraZeneca Foundation
Luc Vermeesch
Vice President Europe, UCB
Kris Westelinck
Managing Director Pfizer Belgium/Luxembourg

WHY HEALTH CARE SYSTEMS ARE NOT PATIENT-CENTERED AND WHAT THIS MEANS

MAX BAUMANN

The first goal of individual health care is to provide suffering human beings (the individual patient) with help and care. That's what doctors and nurses do in their daily work and that's what all of us expect in case of illness – the attention of a helping hand[1]. The target of health care systems – as we have them in most Western countries – is different: it is public health that should be provided to the general public in an appropriate, efficient and economically reasonable way.

Clarifications

All too often, health care politics do not make this basic distinction: to the constituency it is almost every individual's health that is promised; political decisions, however, are decisions about the structures and procedures of and in health care systems.

Health is not something we can 'buy' or even obtain by force. Health is what we lose or do not have when we fall sick or have an accident. Very often, we ourselves cause the loss of health, be it due to bad eating habits (obesity as an example), addiction (alcohol, smoking, drugs) or other risky behavior (e.g. high risk sports). This is why it is incorrect to say that the health care system (or the health care industry) produces health[2]. In the best case, it

---

1   This may also explain why people allocate limited resources to the person who seems to be more sick rather than to another person (appearing less sick, but also sick), even if the possible improvement (for both of them) seems to be equal if measured in Qualys: it's the (more) sick person who primarily needs help and attention.
2   At least not as long as expenditure for prevention is below 3% of the total cost of a health care system. According to the OECD Reviews of Health Care Systems/Switzerland (OECD 2006) the OECD countries spent only an average of 2,7% for prevention (between 0,6% in Italy and a 'peak' of 5,5% in the Netherlands).

helps to restore a lost health, respectively to reduce the consequences of an illness or an accident; these in competition with (at least) two other great healers: the self-healing capacities of human beings and the effect of placebo treatments[3].

From a commercial point of view, the health care industry cannot even have an interest in too many healthy or over-healthy people: sickness is its business, and if everybody was healthy, the whole industry would go bankrupt very quickly. By this, I do not mean to belittle the great achievements of medicine and pharmacology, but to clarify their place and role.

All enterprises within the medical industry function according to the general principles of commercial enterprise: i.e. to make a profit in order to secure the future of the enterprise in a competitive global market. The crux of the matter is, that there is no market – at least not in the ideal form of economic theories – in health care. The 'end-users' (the patients) are normally not in a position where they have a free choice, but are often in existential plight (if suffering from a serious illness). And even if it is widely acknowledged that they should give their informed consent to any medical measure – in practice there is all too often no real information provided that the patient can understand or which he/she can cope with and therefore there is often no real consent at all.

The doctors and nurses, who effectively provide care to patients, are bound by various public and private constraints (e.g. budget restrictions, insurance agreements). Another problem – which should receive more attention – is the over occupation of medical professionals with scientific results, administrative requests and high daily workloads, which also contribute to the fact that they do not qualify as really free players in an alleged health care market.

Health care politics is also the servant of several masters: it is not only about public (not individual) health, but also the financing of the system, the allocation of limited resources to (and within) the system (in competition with other public infrastructures, such as education, security, public transportation, etc.), the protection of local structures (the city hospital) and jobs (in

---

3   Other great healers with proven positive effects are pets.

most Western countries the health care sector is one of the biggest employers; in Switzerland: 12% of all employees), etc.

Quality

The quality of medical interventions should not be assessed in clinical tests alone, but primarily in practice: only what really reaches the patients is helpful (or in some cases harmful) to the end-users. Clinical research therefore needs to be accompanied and completed by implementation research, looking for if, when and how results from clinical research effectively reaches the patient and with what (if any) positive or negative effects.

This is the only reliable way to research the economic merits of a product or treatment. Good clinical results do not necessarily reflect the real effects in practice, which at the end of the day are all that count[4]. And only a coordinated combination of clinical research followed by implementation research and economic analysis of the practical effects can provide a reliable basis for the fair allocation of limited resources. This combination of clinical, practical, economic and ethical research could be labeled as Integrated Evidence Based Medicine, the need and merits of which is indisputable. The reality, however, looks rather disappointing and discouraging:

a) A lot of clinical research is undertaken and millions and millions are spent – but with rather poor results if scrutinized under scientific criteria. It is very sobering to discover – without going into any depth – that the quality standards of around two thirds of all clinical research are questionable[5].

---

4   The results of a German Internet poll that asked the question: "Do you take the medicine as prescribed by your doctor?" showed that only 52,5% of respondents follow the doctor's instructions, whereas over 24% did not take the medicine or stopped taking it, when the symptoms disappeared (13,4%) or took it at their own discretion (9,8%). Not yet considered is the deviation between clinical recommendations and effective instructions given by practitioners. Only the possibility that far less than 50% of clinical results lead to corresponding action in practice make it very clear that implementation research is the next great challenge to be coped with.
5   Suzanne R. Hill/Andrew S. Mitchell/David A. Henry reported in their Review of Submissions to the Australian Pharmaceutical Benefits Scheme that 218 (67%) out of 326 submissions had significant problems and 31 had more than one problem. (JAMA, Vol. 283, April 16, 2000, 2116-2121). This is only one example out of a rapidly growing number of quality assessments that come to similar (or even worse) conclusions.

b) Poor clinical research is one thing; the pitifully few qualified implementation studies are an even bigger problem because (as already mentioned) it is not what has been tested in a clinical research program, but what doctors and nurses finally use and do that is decisive for the health of the patient population.
c) The 'missing link' of qualified implementation research also diminishes the value and validity of economical studies that are based on clinical results alone, not withstanding the questionable quality of such results.
d) Finally it seems almost impossible to make fair decisions concerning the allocation of limited resources if nobody really knows which product and procedures are really of help to suffering patients.

Innovation

Innovation is a powerful motor of change, but not always and not necessarily for the better. If an innovation is a real improvement, this only shows after a certain period of practical experience with the new product or procedure. Innovation is also a 'big seller'. It seems to be an inborn feature of human behavior, to react positively to whatever it is, if it is new[6]. This makes it easy to promote newly developed products of whatever kind to a public greedy for novelties. Informed patients should from their side at least ask the question, if the newest treatment, the newest drug, is really the best and if the novelty value balances the risk of the experimental, not yet proven in practice.

Medical interventions of whatever kind need to be monitored over a sufficiently long period of time because a momentarily successful treatment can show unexpected and detrimental long-term side effects. It is true that clinical research extends over respectable periods. However, even in the best cases it is not able to detect what will happen if a drug or a treatment is used for 10, 20 or even more years. But more importantly, as already mentioned above, clinical studies (in the best case scenario) produce results obtained under optimal controlled clinical conditions. These, however, are very often far removed from what happens in practice.

The positive effects of innovations produced under clinical conditions may

---

6   See Gregory Berns: *Satisfaction. The Science of Finding True Fulfilment,* Henry Hold & Co., New York 2005.

be lost – fully or partially – under practical conditions or even worse, be turned into the very opposite[7]. The lack of comprehensive and reliable implementation research effectively turns even the best medicine based on clinical experience alone into an experimental medicine with open and uncertain effects on patients.

The already high and still accelerating innovation rate, the commercial imperative to promote novelties or perish, turns medicine in practice into a permanent experimental medicine. It becomes more and more unlikely that a new drug or treatment is long enough on the market, and better still used in the daily practice of doctors, to acquire long-term experience like for example Aspirin, which has been used for nearly 100 years.

Transparency

To disclose the goals of a research program in advance and abide to them is not the unquestionable standard it should be. All too often the goals are adjusted during research programs if the obtained results do not comply with expectations, that is, the expectations of the party who finances the program and is eager for quick and commercially usable results.

Unfortunately it is also not a standard to disclose methods and all – also negative – results. Evidence shows that it seems to be common practice to publish (the same) positive results in extenso and under various headings in several scientific papers, but not to publish (or only once, partially and much later) negative results[8]. Not only the results of medical research, but also the costs of developing new products or procedures are obscured rather than made transparent by the providers[9]. In times of exorbitant prices for new

---

7   A good example is the (bad) experience in the fight against tuberculosis in developing countries: medication, which should be taken over a period of many months, is only taken for a few weeks or months, resulting in the bacteria becoming resistant to antituberculotics. It also seems to be common in Western countries that antibiotics, for example, are only taken until symptoms disappear and not until the end of the treatment period (see also footnote 4).
8   Hans-Hermann Dubben and Hans-Peter Beck-Bornholdt at the Universitätsklinikum Hamburg-Eppendorf have carried out basic research concerning the publication bias of medical studies. See for example their report *"Unausgewogene Berichterstattung in der medizinischen Wissenschaft" (Publication Bias in Medical Science)* Hamburg, 2004.
9   See Marcia Angell, former chief editor of the New England Journal of Medicine: *The Truth About the Drug Companies – How They Deceive Us and What to Do About it*, Random House, 2004.

drugs and the limited resources of all Western health care systems, transparent costs become a major issue.

It is obvious that non-disclosure – of results or calculations – may be considered as a protection against and an advantage over competitors. On the other hand, the disclosure of such practices – especially if 'hidden' results lead to damages to patients – may cause serious problems to the industry. Therefore it is in the industry's interest to agree on quality standards for the design, realization, monitoring and publication of research programs and studies. This would not hinder competition if the same terms applied to all competitors. It is also very clear that states or supra-national entities (like the European Union) will promote such standards in their legislation if the industry is unable or unwilling to remedy this unsatisfactory situation.

The same goes for better transparency of costs. Auditing rules in most states tend to oblige enterprises to publish data, which allow the public (especially the share-holders as owners) a true and fair view of a corporation's commercial and financial situation. The available data for the real cost of medical research in most cases do not comply with such a true and fair view-standard.

It is easy to understand, that better transparency as to the (real) merits of medical research will have a positive effect on the quality of the result. In addition, it may also be expected that products and procedures that today pass the requested tests, would no longer do so. Higher requirements as to the quality of research would probably also slow down the innovation rate which, however, must not be a disadvantage as the actual experimental medicine could turn to a more evidence based science. More transparency concerning the real costs of research might have an impact on prices and margins (we will return immediately to this point, below).

Health for money – money for health

Health care spending has been increasing at a faster rate than the cost of living indexes in most Western countries. The gap between increasing costs on one side and a diminishing ability or willingness to pay on the other is widening. At the same time, OECD reports show that higher spending does not necessarily lead to a higher quality of public health care.

The problem is further obscured by the way funds are distributed through complicated and opaque channels, with many dead ends and also leaks (meaning the wrong or inefficient allocation of scarce funds).

We can also see, that certain new services or drugs are offered at prices that have nothing to do with the provider's cost (including a generous margin) but only with peoples' willingness to pay, in other words, seriously ill people who have no other choice. It is clear that under such circumstances, the suspicion of unfair profiteering surfaces: the law in most countries contains in one form or another, provisions that ostracize the unfair taking of advantage of another's distress[10].

It is perhaps only a question of time until certain price fixing policies of the health care industry are scrutinized in this respect. In any case the pretension of the industry is that what they offer is good for the recipient – so good that they can ask almost any price for it. This also makes it – on the other hand – very clear and confirms that the first goal of the industry is to maximize its profit, regardless of moral concerns, as long as no correction – either by the industry itself or by the law – is made. It is also obvious, that this commercially motivated and 'legitimized' practice has nothing to do with helping individual suffering human beings, to whom medicine in the Hippocratic tradition is or should be primarily dedicated.

That is why politicians are asked to install some necessary checks and balances: it is very unlikely that only a market-driven health care system will improve public health. Growth in this market is not achievable with more healthy people but only with more people suffering from more and worse illnesses, which in many cases is not a question of individual wellbeing but of a clinical (industrial) definition of threshold values or refined diagnostics.

Unfortunately it is not only the novelty temptation alone but also the still very large and uncritical goodwill of patients towards the providers of medi-

---

10 An obvious disparity between the respective considerations, whereby one of the parties is exploiting the distress of the other, may allow the latter to rescind contracts (art. 21 of the Swiss Contract Law/Code of Obligations). Criminal Law (e.g. art. 158 of the Swiss Criminal Code) provides up to 5 years imprisonment (in the case of commercial profiteering up to 10 years) for abusing the distress of another person if the counter-performance is commercially obviously disproportionate.

cal services, which in the case of the individual doctor/patient relationship is probably justified in most cases, but which is of doubt with regard to the health care system the more it turns into a system that follows only commercial standards.

As the individual providers (doctors and nurses) work more and more under the constraints of such systems, they face more and more situations of having to choose between the best individual service to their patients on one hand and the most economically reasonable standard treatment, as described in a standard catalogue of diagnostics (DRGs-tariffs [11]) on the other.

[150] Global budgets allocated to the individual institutions in the health care system are an efficient way for politicians to avoid the crucial question of how to allocate limited resources to the individual patient. However, it is meanwhile a common situation within hospitals 'living' from a global budget that the heads of the various departments fight for their share of such a budget in a series of never ending meetings [12].

The moral of the tale

Medical services and products are special and not to be compared to a commodity like clothes or cars. Still they are delivered more and more by enterprises that follow the same logics as clothing or car manufacturers [13]. Medical products and services (life style drugs and a large part of beauty surgery excluded) are not things we normally long to have, but something we need in order to (hopefully) regain (or improve at least) lost health.

Therefore patients, i.e., really informed and autonomous patients, should more thoroughly question services and products and why they are offered to them. Is there really reliable evidence that the proposed treatment is efficient and not even noxious? On the other hand, all of us are potential pa-

---

11 DRG = Diagnostics Related Groups. Meanwhile a whole family of DRG has been born, as e.g. AP-DRG (AP = All Patient), APR-DRG (All Patient Refined), IAR-DRG (International All Patient), IR-DRG (International Refined), etc.
12 To give an example: How should the heads of the other departments react when the (ever increasing) cost of medication for cancer patients (5% of all patients treated) consumes 25-30% of the total available budget?
13 There are still a lot of non-profit players in Western health care systems, which are however struggling more and more with financial restraints.

tients and could minimize the respective risks by avoiding unhealthy behavior (smoking, eating too much of the wrong food, etc.). Responsibility for one's own health [14] is a concept that should be more strongly promoted, together with the dissemination of more information about and insight into the functioning of health care systems, their products and services.

The latter, however, requires transparent, reliable and comparable information standards for all players in the game: i.e. insurers (private and public), the medical industry and service providers and – last but not least – patients. Correct and comprehensive information could also improve the present unsatisfactory quality of research and the publication of research results. Here also, those concerned should develop general quality standards concerning research programs, financing of such programs and the publication of results.

For their part, politicians should make efforts concerning the transparency of their own intentions and relations [15]. They should also promote the transparency and availability of data because they and everybody else need to come to educated decisions (or at least educated guesses). Further, they should closely monitor and accompany the elaboration of the aforementioned standards and – if the providers are unwilling or unable to improve the actual situation – to enact the necessary rules by way of legislation.

The most crucial and unpopular decisions politicians should take and have to take are those concerning the allocation of scarce funds. Global budgets are certainly an easy way to avoid such critical problems, but they are in no way appropriate. They lead to unfair allocations and only shift the burden of crucial decisions to institutions, doctors and nurses. It is unacceptable that the treatment of a patient should depend on how many other similar cases have already been treated in the same institution. Nor is it acceptable that so much energy is wasted in a battle over budget share between institutions and even within institutions (see the example in footnote 12).

---

14 Before medicine could really help it was of vital importance to take care of one's own health according to the guidelines of the six canons of classical dietetics: aer/cibus et potus/motus et quies/somnum et vigilia/excrementa et secreta/affectus animi (clean air/eating and drinking/action and rest/sleep and being awake/excretion and secretion/emotions).
15 It should at least become more visible as to whether decisions concerning the health care system are aimed at improvements for patients or securing local structures in the respective politician's constituency.

On the other hand – if improvements to transparency and quality of services and products are carried out – it may be assumed that available resources could easily cover every requirement, proven by clinical efficacy, effectiveness in the practical application and economic efficiency.

Patients would not receive everything, but everything they did receive would be backed up by a quality report according to the standards of the proposed Integrated Evidence Based Medicine. It is certainly true that catalogues always contain a certain risk; that the available (listed) services or products do not really comply with the needs of an individual patient. However, if such catalogues are based on seriously researched results, this risk will diminish. And it would also be possible to grant doctors a sort of 'emergency reserve' from which they can draw at their discretion in order to solve the remaining 'hard cases'. However, the ordinary day-to-day decisions as to which services are provided by a public health system, should be made on the basis and result of public debate.

Conclusion

A Western health care system is a game with many players amongst whom the patient has only recently been detected as an active subject and not a mere recipient of health care services. As tax payers and debtors of ever-increasing premium payments, we are all interested in keeping health care spending as low as possible; but a lot of people are also living from the health care industry (e.g. as one of the biggest employers, offering interesting and safe jobs). And if we fall sick, we expect the best treatment available. However, not to become ill or even dependent due to bad health is something we all strive for. We do not like to see ourselves in the role of a patient and this is not a central perspective as long as we are not severely ill.

That is why we – in other words, all the players in the game: patients, physicians, providers of services and financing – need an increased sense of responsibility in our various and varying roles. A responsibility for our individual as well as public health will not arise automatically if we only optimize our systems from an economic point of view; the 'invisible hand' alone will not solve all of our problems, we have to lend a hand ourselves. In short: what we need is more moral awareness.

BIO

Max Baumann (born 1951) is cofounder and member of the Board of Dialog Ethik (Interdisciplinary Institute for Health Care Ethics/Zurich) and Member of the Swiss Federal Commission for Basic Issues of the Health Insurance System. He is an Attorney-at-Law, Dr Iur., and Professor of Philosophy/Theory of Law and Civil Law (University of Zurich). His latest publications concerning health care issues are "Vorsorgeauftrag für medizinische Massnahmen und Patientenverfügung" (ZVW 2005/2, 58ff) and "Recht?Ethik?Medizin/Eine Einführung ins juristische Denken – nicht nur für Ethiker und Mediziner", Peter Lang Verlag, Bern etc., 2005.

PHYSICIANS AND THE PHARMACEUTICAL INDUSTRY: A SENSITIVE ISSUE

MARC BOGAERT

The interaction between physicians and the pharmaceutical industry has in recent years been a topic for discussion all over the world. There have been many publications on this subject [1-4], often in outstanding medical journals, but also in the lay press. This issue is also of interest for Belgium, and attention was already drawn to it in 2002 by the Belgian academies of medicine [5]. I have chosen to restrict the scope of my text to the interaction between physicians and the pharmaceutical industry, but the problem also exists for the contacts between, on the one hand, other health professionals, hospitals, decision-makers, patient organizations, journalists, and on the other, all industries active in the field of health care. Physicians have contacts with pharmaceutical companies in very different settings. Although the distinction is somewhat artificial, for the sake of clarity I shall discuss separately the physician as (1) researcher, (2) expert for the decision-makers, (3) teacher and opinion leader, and (4) prescriber. I will end with a few recommendations for the different stakeholders.

The physician as researcher

Many physicians participate in industry-sponsored studies, and it is important that such studies continue to take place in Belgium and that our clinicians are involved in them. However, it is my experience and the experience of others active in ethics committees that clinicians participating in commercial clinical trials are often not involved in the elaboration of the protocol nor in the evaluation of the results; they simply provide patients. They are also not interested enough in the relevance of the study in which they collaborate. The decision as to whether to publish the results of a study is usually left to the discretion of the pharmaceutical sponsor. Results of a clinical trial should be made public, and no restrictions can be accepted, although for commercial reasons a delay in publication may be warranted. It is to be deplored that neither the European Clinical Trial Directive (2001/20/EC) nor the Belgian law on Experiments on Humans of 7 May 2004 mentions the publication of clinical trial results. In the Netherlands from 1 January 2009 onwards, ethics committees can no longer accept a protocol if agreements about publication are not clearly mentioned[6]. It is obvious that the publication should reflect the results correctly, whereas this is not always the case[7]. In the last few years there has been growing interest in ghostwriting the results of clinical trials for publication[8-10]. And I do not exclude the fact that sometimes the authors of a publication concerning a clinical trial have not seen the text before it is published.

Nowadays, most journals request that any conflicts of interest on the part of the authors should be clearly stated, and this brings me to the financial arrangements between the physician and the pharmaceutical sponsor. Both the European Trial Directive and the Belgian law require that the ethics committee should evaluate all financial agreements. The physician's fee (as well as financial agreements with the hospital) should be proportional to the services rendered. Lack of proportionality can lead to minor or major violations of the protocol, sometimes to outright fraud[11]. It is not known how closely our ethics committees, as spelled out in the Directive and the Belgian law, look at financial arrangements, and what criteria they have set up to judge whether the payments proposed are acceptable.

Finally, I have the impression that, perhaps due to the requirements of the Directive, academic research is declining: it is becoming much more difficult for clinical researchers to set up their own trials, rather than to participate in a commercially sponsored trial. Academic research remains, however, essential[12].

The physician as expert for the decision-makers

Decision-makers everywhere rightly insist upon transparency about possible conflicts of interest. The fact that it is very difficult to clearly define what in a given setting constitutes a conflict of interest should not be used as an excuse to ignore this issue. In this regard it is important to mention that a link with a pharmaceutical company can not only constitute a conflict of interest for the expert when the product of that company is discussed, but also when the expert participates in discussions about the product of another company which markets competing drugs. The Belgian advisory committees (e.g. on drug registration and drug reimbursement) each have their own requirements concerning the declaration of conflicts of interest of their members. These declarations are archived, but it is not clear to me how these committees (or the authorities) evaluate them and control their veracity, and how they decide whether in some cases the declarations should lead to the decision not to accept somebody as an expert for the committee. I am fully aware that, certainly in a small country, it can be difficult to find experts for a particular problem who do not have links with the company involved or with competing companies.

The physician as teacher and opinion leader

Many physicians from academic institutions are involved in graduate and post-graduate teaching and training. There has been interesting commentary on this, for example, a paper which concludes: *"Overall, institutional academic-industry relationships are highly prevalent and underscore the need for their active disclosure and management"*[13]. In the last few years, academic institutions have tried to set up their own code of conduct for the relationship between their teachers and the pharmaceutical industry[14]. In June 2008, the American Association of Medical Colleges issued a report that contains *"recommendations about gifts to individuals, pharmaceutical samples, site access by pharmaceutical and device manufacturer representatives, continuing medical education, participation in industry-sponsored programs, industry-sponsored scholarships and other educational funds for trainees, food, professional travel, ghostwriting, purchasing..."*[15].

More and more physicians from academic institutions and from scientific and professional organizations are involved in the elaboration of guidelines, and give their opinion about medicines in publications and presentations at symposia and congresses. The fact that physicians involved in the elaboration of guidelines, often have manifest conflicts of interest has been criticized[16]. More and more, the need for transparency about these conflicts of interest is stressed. Editors of scientific journals have a special responsibility. In the British Medical Journal, Horton, the editor of the Lancet, was cited as follows: *"The relationship between medical journals and the drug industry is somewhere between symbiotic and parasitic, according to the editor of the Lancet,*

Richard Horton. But at the moment it has swung too much towards the parasitic, he told the House of Commons select committee on health last month in his oral evidence on the role of the industry"[17]. Last year the discussion about drug advertising in medical journals was the topic of a yes/no debate in the British Medical Journal[18]. Editors of the so-called 'independent' drug information journals (in Belgium, these include the Folia Pharmacotherapeutica and Minerva) should be even more attentive to this problem.

The British Medical Journal of 21 June 2008 contains several papers about opinion leaders. The cover features the picture of a physician as a puppet on a string accompanied by the headline "Should the drug industry use key opinion leaders?". The main question asked is "Key opinion leaders: independent experts or drug representatives in disguise?"[19]; there is also a yes/no debate on "Should the drug industry use key opinion leaders?"[20], and the editor of the journal admonishes: "Key opinion leaders, your time is up"[21].

In this regard, it is worth mentioning an anecdote about an NIH-conference planned in 2007 on guidelines for screening pregnant women for herpes. A number of distinguished scientists, amongst them Kassirer and Angell (former editors of the New England Journal of Medicine), Horton (editor of the Lancet) and the head of the Cochrane Collaboration, signed a letter protesting that four out of five of the speakers for the conference had undisclosed ties to pharmaceutical companies that made antiviral drugs[22]. The conference was cancelled.

The physician as prescriber

Rational prescribing begins with correct decisions by the authorities about registration (and the Summary of Product Characteristics) and reimbursement (and the criteria surrounding it). Graduate and post-graduate training in this should be appropriate. The importance of continuing medical education is obvious: the pharmaceutical companies should not influence initiatives leading to accreditation, but in Belgium this is not always the case.

Pharmaceutical companies try to promote their products through advertising, for example, in medical journals. In 2007 the Belgian 'Kenniscentrum' (KCE) published a report on the evidence base of written information sent by the pharmaceutical industry to general practitioners: only a minority (17%)

of the messages had a real scientific base or were based on the Summary of Product Characteristics[23]. In some periodicals sent free of charge to physicians, the majority of messages pertain to medicines; as these periodicals are totally financed by the pharmaceutical industry, it is not surprising that it is often very difficult to see the difference between acknowledged publicity, and scientific messages, given by opinion leaders.

What about the incentives for the prescriber? The conclusion of a paper on *"Physicians and the pharmaceutical industry. Is a gift ever just a gift?"* was as follows: *"The present extent of physician-industry interactions appears to affect prescribing and professional behavior and should be further addressed at the level of policy and education"*[24]. This conclusion is certainly still valid in 2008. Statements such as *"small gifts do not influence the prescriber"* and *"I am not influenced by gift"* are myths. Attempts have been made to define the dividing line between 'acceptable' and 'not acceptable' incentives. These attempts are largely unsuccessful, and perhaps a 'zero tolerance' policy is the only solution. In the meantime, greater transparency is needed.

There is, understandably, often pressure from patients on the prescriber. In the last few years, there has been much interest in the links of patient organizations with pharmaceutical companies[25-26]. Umbrella organizations such as the 'Ligue des Usagers des Services de Santé' (LUSS) and the 'Vlaams Patiëntenplatform' (VPP) play an important role in that regard. The Belgian Association of Pharmaceutical Industries (Pharma.be) has recently included in its code of deontology, rules about the relationship of their members with patient organizations, including conditions, transparency, and guarantee of the independence of the organizations.

A few recommendations

There will always be (and should be) contacts between physicians and the pharmaceutical industry. Transparency is essential, but is not enough. There should be regulations, monitoring, and where needed, direct action. But who should act?

The legislator has already formulated laws and regulations for the industry, but also for physicians: there are obligations for both parties. For example, there is the European Directive 2001/83/IC (modified by the Directives 2004/24/EC and 2004/27/EC), the Belgian law on medicines of 24 March 1964, the Royal Decree 78, the Royal Decree of 23 November 2006 concerning Mdeon (control of hospitality offered by pharmaceutical companies). In my opinion there are enough rules and regulations, but the authorities should now set up the necessary structures to monitor what is happening in the field, and to intervene when necessary. It is perhaps worthwhile to harmonize the procedures for obtaining the declarations of interest in the different advisory committees, and to set up an independent body whose task would be to evaluate these declarations and take action where necessary. Those who are responsible for the accreditation of physicians should make sure that the continuing medical education initiatives are free from influence from industry.

The ethics committees should, when evaluating clinical trial protocols, pay more attention to the relevance of the study, and to the financial arrangements between all parties involved, tasks set out clearly by the European Directive and the Belgian law. They should evaluate the arrangements about the publication of results.

Medical schools and university hospitals should introduce a code of conduct for their physicians and take the lead in assuring transparency with regard to the relationship between health professionals and the industry.

The pharmaceutical industry association (Pharma.be) has its own code. The question should be asked as to how far its members adhere to it. More transparency, and publication of violations are needed. The pharmaceutical industry should have the courage to admit that violations occur and should be encouraged to resist undue pressure from the medical profession.

Physicians and their organizations should refrain from ignoring the problem, and should avoid placing undue pressure on pharmaceutical companies. The 'Ordre des Médecins' is rather vague in its code, but it has given support to the advice of the Academies of Medicine of 28 September 2002[27]. Professional and scientific organizations of physicians should take initiatives to raise their members' awareness on the interaction with the pharmaceutical industry; they should set up codes of conduct and should fully endorse initiatives that are taken to enhance transparency.

REFERENCES

1. Angell, Marcia. *The truth about the drug companies – How they deceive us and what to do about it,* 1st edition. Random House, 2004.
2. Campbell EG, Gruen RL, Mountford J, Miller LG, Cleary PD, Blumenthal D. *A national survey of physician-industry relationships.* New Engl J Med. 2007;356:1742-50.
3. Campbell EG. *Doctors and drug companies – scrutinizing influential relationships.* New Engl J Med. 2007;357:1796-7.
4. Kassirer JP. *Financial conflicts in the medical profession: an ongoing, unresolved problem.* Open Medicine 2007;1:123-5.
5. *Advies van de Belgische academies voor geneeskunde over de relatie tussen artsen en de farmaceutische bedrijven* (28.09.02); www.academiegeneeskunde.be
6. CCMO press release (13.11.08); www.ccmo.nl
7. Chan A-W, Hróbjartsson A, Haahr MT, Gøtzsche PC, Altman DG. *Empirical evidence for selective reporting of outcomes in randomized trials.* JAMA. 2004;291:2457-65.
8. Ross JS, Hill KP, Egilman DS, Krumholz HM. *Guest authorship and ghostwriting in publications related to rofecoxib.* JAMA. 2008;299:1800-12.
9. Gøtsche PC, Hróbjartsson A, Johansen HK, Haahr MT, Altman DG. *Ghost authorship in industry-initiated randomised trials.* PLoS Medicine. 2007;4:e19.
10. Sismondo S. *Ghost management: how much of the medical literature is shaped behind the scenes by the pharmaceutical industry?* PLoS Medicine 2007;4:e286.
11. Ross DB. *The FDA and the case of Ketek.* New Engl J Med. 2007;356:1601-4
12. Remuzzi G, Schieppati A, Boissel J-P, Garattini S, Horton R. *Independent clinical research in Europe.* The Lancet. 2004;364:1723-6.
13. Campbell EG, Weissman JS, Ehringhaus S, Rao SR, Moy B, Feibelmann S, Goold SD. *Institutional academic-industry relationships.* JAMA 2007;298:1779-86.
14. Brennan TA, Rothman DJ, Blank L, Blumenthal D, Chimonas SC, Cohen JJ, Goldman J, Kassirer JP, Kimball H, Naughton J, Smelser N. *Health industry practices that create conflicts of interest. A policy proposal for academic medical centers.* JAMA. 2006;295:429-33.
15. Association of American Medical Colleges. *Industry Funding of Medical Education.* Report of an AAMC Task Force (2008); www.aamc.org/publications
16. Steinbrook R. *Guidance for guidelines.* New Engl J Med 2007;356:331-3.
17. Eaton L. *Editor claims drug companies have a "parasitic" relationship with journals.* BMJ 2005;330:9
18. Smith R/Williams G. *Should medical journals carry drug advertising? Yes/no.* BMJ. 2007;335:74-5.

19. Moynihan R. Key opinion leaders. *Independent experts or drug representatives in disguise?* BMJ. 2008;336:1402-3.
20. Buckwell C/Fava GA. *Should the drug industry use key opinion leaders? Yes/no.* BMJ. 2008;336:1404-5.
21. Godlee F. *Editor's Choice – Key opinion leaders, your time is up.* BMJ. 2008;336:1384.
22. Tanne JH. *Group asks US institutes to reveal industry ties.* BMJ. 2007;334:115.
23. Report of the Belgian KCE (Kenniscentrum). *Geschreven informatie van de farma sector aan artsen: vooral marketing en minder wetenschap* (29.06.07); www.KCE.fgov.be
24. Wazana A. *Physicians and the pharmaceutical industry. Is a gift ever just a gift?* JAMA. 2000;283:373-80.
25. Kent A./Mintzes B. *Should patient groups accept money from drug companies? Yes/no.* BMJ. 2007;334:934-935.
26. Chalmers I. *The Alzheimer's Society, drug firms, and public trust.* BMJ, 2007;335:400.
27. Orde der Geneesheren. *Advies van de Nationale Raad. Relatie artsen met farmaceutische industrie – Gemeenschappelijk advies van de Koninklijke Academies voor Geneeskunde van België* (16.11.02); http://195.234.184.64/frame-totaal.htm

BIO

Marc Bogaert is a clinical pharmacologist, specialist in internal medicine and holds a PhD in pharmacology. He was professor of pharmacotherapy at the Heymans Institute of Pharmacology at Ghent University (Belgium) until he retired in 2000. He was dean of the Ghent Medical School from 1981 until 1994, and permanent secretary of the Academy of Medicine from 1994 until 2004. He was editor-in-chief of the publications of the Belgian Centre for Pharmacotherapeutic Information until 2007.

PART THREE
NEW FRONTIERS

COMMON GOALS TO GUARANTEE ACCESSIBLE HEALTH CARE: A POTENTIAL WIN-WIN SITUATION OR JUST A (BAD) DREAM?

GUY PEETERS

"Corporate governance is about a much larger group than the shareholders of the pharmaceutical industry. We're talking here about the governance of all the stakeholders: doctors, pharmacists, hospitals, mutualities, policy makers."

The very broadly and generally formulated subjects that are supplied by CROSSTALKS possess the advantage that they leave plenty of room for maneuver.

In my case, I could have restricted myself to a discussion with a high consultancy level and an equally high concentration of hot air about the 'business model' corporate governance that is to be applied in health care. Or, I could have tried to venture a more conceptual academic approach. In this line of thinking, I would have demonstrated the theoretical advantages that can lead to a corporate governance attitude of the different actors in the system. I would have looked for the necessary preconditions that the respective actors should comply with in order to achieve such a conceptual surrounding. No need to add that this conceptual academic approach starts from a tabula rasa scenario.

With almost 25 years of experience as an observer and sometimes as an active player in the decision-making process of the Belgian health care at the back of my mind, it seems more appropriate to start from the existing situation and verify from a pragmatic point of view what concrete and realistic steps can be taken in order to come a bit closer, from a conceptual point of view, to an ideal corporate governance attitude by all the players.

Such an approach is the closest to reality: in Belgian health care, we never start revolutions, but always work step by step and always too slowly according to some. Therefore: evolution and not revolution.

In the first place, my discussion will therefore focus upon a description of the situation of Belgian health care in the medium-term, emphasizing the challenges that we will be confronted with. Secondly, I will explain how the pragmatic use of concepts of corporate governance in the health care sector in general and in the pharmaceutical sector in particular, may make that the affordability, the accessibility and the quality of our health care can be rendered permanent and be protected for the coming generations.

The future context of Belgian health care

Just like in the other European countries that finance their health care system mainly by public finances, either coming from general means or by means of social security contributions (or through a mixture of both), Belgian health care and by extension the West-European traditional social model will come under financial pressure in years to come. Two evolutions are at the basis of this: on the one hand, the encroaching ageing of the population and on the other, globalization and the international fiscal context. It becomes more and more difficult for individual countries to make full use of the financial instruments in order to achieve balance in public finance.

With the introduction of the EMU (European Monetary Union), monetary instruments already nearly completely disappeared and with globalization, the space for fiscal maneuver in order to generate additional income for the government is very limited. This implies that for the government, choices must be made within inflexible budgets in order to finance challenges such as the ageing population.

For that matter, we find to our great regret that the period of non-policy and its very negative consequences for the global federal budget we are observing in Belgium since the last federal elections of June 2007, will restrain the financial margins in the health care sector. It seems that Belgian politicians of this generation have made the same errors as those of the seventies...

Moreover, Belgium – unlike most other European countries – is also confronted with two elements that even strengthen this internationally created financial frame of reference.

First of all, there is our historic national debt, which now represents 85% of GDP (2007) but which should drop to 60% in order to comply with the European stability pact. Secondly, we have a state structure that places almost the entire financial burden of the (future) ageing population in terms of social security (and mainly in pensions and health care) firmly on the shoulders of the federal government, which is already short of money.

As a matter of fact, these two typically Belgian phenomena – the important national debt and federalism – are historically linked. When in 1980 the net balance to be financed increased spectacularly in six months from 6% to 13%, this was for a great part due to the fact that between the collapse of the Egmont pact and the overthrow of the Eyskens government, the political class had mainly put its energy into resolving institutional quarrels instead of logically tackling the enormous economic problems at that time.

Finally, in the last couple of years EU Member States are more and more confronted with a relatively recent phenomenon, namely the growing horizontal test by the European Court of Justice (ECJ) of the basic principles of the European internal market against the organization of their respective social security systems. This limits their space for maneuver even more in the field of policy with regard to social protection and health care in particular.

Even though the EC Treaty explicitly excludes national social security systems from the functioning of the internal market, the constant jurisdiction of the ECJ has made it clear since the decree Decker/Kohll of April 1998 that Member States can no longer do whatever they want in order to offer their citizens an acceptable standard of health care at an affordable social price. Even though many Member States are resentful and irritated about what is considered as excessive interference by the European Commission in internal affairs, it is clear that as actors within the health care, in years to come we will have to learn to live with the Commission's and ECJ's 'right of injunction' in this matter.

The public statement by an important person in charge of the GD Internal Market, Mr. Karel Van Hulle, at the end of 2007 during the colloquium European Integration and Health Care Systems at the NISDI[1], that *"when Member States call upon the 'general good' principle, this is often just an excuse to maintain protectionist measures"*, is in this respect clear and explicit enough.

Finally, it is certain that new Member States, those which joined since 2004, are favorably disposed towards a liberal Anglo-Saxon health care model – partly because this costs their poverty stricken public finances little or nothing – rather than the Rhineland model, which the old Member States who are slowly getting into hot water, have always defended.

In a nutshell, we can therefore say that Belgian health care will be confronted with the following challenges in the coming years:

1. The seven wealthy years for sickness insurance are clearly over: in spite of the expected surplus expenditures, it appears from all statements by the new coalition that the real effective growth rate will be significantly reduced. Even if theoretically the authorized legal 4,5% growth in *real terms* is maintained for 2009, it is clear that a large part of this growth rate will be reserved for future ageing expenses. What is far less clear is what this will cost the patient in terms of increased personal contributions.

2. Organizationally, the structural switch from cure to care in our health care system is yet to be started: the development of (cheaper) alternative housing does not get off the ground, so what are we going to do with the additional 72,000 dementia patients we will be confronted with by 2020 compared to 2000? I know, this may seem to be a conflict of interests, but I must also think of my own future. Have any structural measures been taken in the field of care functions in professional training? Where are the promotional campaigns about the possibilities and advantages of such training, aimed at specific target groups, such as immigrants and semi- and unskilled workers?

---

1 National Institute of Sickness and Disablity Insurance

3. Finally, the government – be it the federated or the federal – and all the actors must realize that supranational European control will only get worse. Therefore, it's no longer any use to try to come to a decision sneakily. Big Brother is and will be watching us and will continue to do so in an even more efficient way.

In this pessimistic scenario, the future affordability and accessibility of the health care system is not obvious.

It will be necessary to use all the available means as efficiently as possible. One of the ways that might irrefutably lead to this, is if the government could create an environment in which the individual interests of the different actors within the health care sector could run in parallel and ideally even converge, and this to the advantage of the patient. This could minimize inefficiencies and waste of means due to conflicts of interest.

Speaking learnedly, these dynamics could be described as corporate governance.

Corporate governance in Belgian health care: A pragmatic approach

Corporate governance applied to health insurance could be described as the joint ambition of all active actors (practitioners, industry and health funds) to guarantee the passive actor, the patient, a form of health care that is as qualitative and accessible as possible and this within the fixed budgetary means that society is willing to cough up for this. In this model, the government could play a supporting financial and regulating role.

In order to achieve this, the actors, in an atmosphere of maximum transparency, should come to a joint long-term vision about the way the allotted budget should be spent in the most efficient way, based upon the most recent international scientific consensus and therapeutic state of the art.

Saying that this definition of corporate governance would be totally alien within Belgian health insurance would seriously bend the truth and the daily efforts of the thousand practitioners in this country in favor of their patients.

It can be said without much exaggeration that the Leburton Act of 1963, which gave shape to health insurance as we know it today, was in the spirit of that time certainly a legal translation of the corporate governance principles that have only caught on in the business world during the last 10 to 15 years. In exchange for social rates guaranteeing the patient an accessible health care, practitioners and health funds have the opportunity to establish their priorities by means of a negotiation process, in which the government implicitly guarantees the income of these practitioners.

The Moureaux Act of 1993 has continued in this direction and has actually laid down the principle of the separation of the roles of the supervisory bodies and the operational bodies on the one hand and the principle of the accountability of the different actors, in particular of the health funds, on the other. After all, it is now up to the General Board – in which financiers have the final word (government and social partners) – to establish the general lines of policy and budget. The Insurance Committee, with an equal representation of health funds and practitioners, is charged with the operational translation of these decisions.

Finally, I will mention the Busquin Act of 1990, which has caused necessary shocks in the internal functioning of the health funds. The Busquin Act introduced the rules of internal democracy and – at least equally important – enforced financial transparency and external and internal control. The impact of this legislation on the preservation and strengthening of the social credibility of the health funds world cannot be emphasized enough. I believe I can say that, for a great part, the health funds owe their survival to the fundamental reforms brought by the Busquin Act.

For my organization, it is thus beyond dispute that the further the principles of corporate governance are extrapolated within a firm or within a sector – especially when financed by public means – the more favorable the consequences will be for that firm (and its shareholders), the sector and society in the long-term.

Do I hereby mean to declare that everything worth saying about corporate governance has already been said in the health insurance sector? Of course not, and in my opinion this is due to two fundamental aspects: on the one hand the intense human interactions typical of activities in health care and on the other, the obvious and deliberate legal denial – until recently – of the existence of one of the leading actors in the sector, namely the pharmaceutical industry.

Concerning the first point, human interactions, I mean two basic elements that represent for a great part the strength of our system. In the first place the general consensus about the almost absolute respect of what we call the *colloque singulier* between patient and practitioner, which constitutes the foundation of a relationship based on mutual trust. Secondly, I refer to the fee-for-service financing system that mainly characterizes our system and which is at the basis of the (financial) enthusiasm of our practitioners.

These two elements not only represent the strength of our system, but also and at the same time a fundamental weakness as far as corporate governance is concerned. Very often – and usually this behavior can in one way or another be legitimate – they make practitioners focus on the individual interests of the patient or – which is more difficult to justify, but perfectly understandable – on their own financial situation rather than on the general interest, which in this context may be translated as a correct and rational use of public funds in health insurance.

Two small examples

An example from the end of the nineties: Hytrin®, an alphablocking drug from the firm Abbot, was registered as an anti-hypertension drug and a drug against benign prostatic hypertrophy. As an anti-hypertension drug, Hytrin® was reimbursed, but when prescribed for benign prostatic hypertrophy, patients had to pay from their own pocket.

Although the treatment of hypertension by alphablocking agents was at that time already very controversial, the turnover of Hytrin® on the Belgian market certainly did not appear marginal. However, what became clear when looking at the patients' profiles who were prescribed Hytrin®, was that 97% of them were male and of these 97%, 95% were 50 years or older. We also found that an important percentage of the prescribers were urologists… Confronted with this, it soon became clear that doctors confuse therapeutic freedom of prescription with a correct application of the regulation.

A second example: imagine a hospital situated in a region with few rosy socioeconomic prospects and a high rate of unemployment. On top of this, it is the largest employer in the region, and has difficulty charging supplements because of the socioeconomic profile of the patients. Isn't it understandable in this context that the hospital is guilty of systematic overconsumption in order to guarantee the financial equilibrium and employment it represents for the region? It probably is. And can we call this a rational use of means? I think not.

Such inefficiencies will probably be curtailed by the introduction of clear rules, (an even) better and faster monitoring of and real sanction possibilities against practitioners who systematically overstep the mark. However, in this debate we must avoid walking into the trap of overregulation, in the name of the holy cow 'efficiency'. Today, we are confronted with a situation where for every €100 invested in the compulsory health care system, €95 go to the patient. Of these €95, two or three are likely to be spent inefficiently. However, all in all, to me this still seems to be much more favorable for the patient than the American or, closer to home, the Dutch situations, where the part of the global budget that goes to the patient is perhaps spent more efficiently, but where administrative costs of 15% up to 20% are necessary in order to achieve this rate of efficiency. So, *in fine*, what's better for the patient: a small fraud level with small operational costs and high health care coverage or no fraud with high operational costs and smaller health care coverage?

Finally, many attribute these forms of inefficiency to the widespread vertical and horizontal forms of conflict of interests that are ingrained in the whole system and therefore plead for a total and absolute separation between the field and the decision-makers. However, such a move would very soon have catastrophic and incalculable consequences that would surely affect the patient. We would soon find ourselves in the same situation as that of the newly graduated master of agricultural science who, gazing at the farmer's, says: *"You won't harvest many apples this year"*, whereupon the farmer replies: *"No, certainly not with these pear trees"*. The complexity of the sector and its continuous evolution mean that policy needs almost daily concrete input and follow-up from the field of action. Therefore, we need links that are active on both levels and that can, according to circumstances, (simultaneously) wear different hats (at the same time).

Corporate governance policy in this field, and in particular as far as transparency is concerned, would then consist of clearly pointing out in what quality these links intervene, whilst making clear to the other parties concerned in a specific case what interests these links represent.

So I have always found it strange that the advice of university experts, for example in the former Technical Board for Pharmaceutical Specialties – which had to decide on the reimbursement level of new drugs – was never published at the time, as if one liked to offer these experts the possibility of taking their positions, on which they would otherwise be judged without mercy in the academic scientific milieu, in closed cenacles. This is just my impression, not a certain or proven fact.

This example brings us to the second fundamental point that has to be tackled in order to optimize and strengthen the corporate governance mechanisms in health insurance. It concerns the mutual and explicit recognition of the government and the pharmaceutical industry as stakeholders. The government must aspire to recognize the industry as a full actor and treat it as such, with specific interests, but also with an explicit added value for the sector and the patient. The industry and their shareholders must realize that in their market, to a large part financed by public means, the government is more than just an ordinary client, but often the reference stakeholder who in the development of a lasting relationship based on mutual trust can guarantee in the long-term the basic return that their shareholders expect in exchange for an affordable innovation. But how can this vision be concretely achieved?

Towards a new government-industry relationship based on corporate governance principles

A strategy of confrontation was the rule up to eight years ago when relations between industry on the one hand and government and the health funds on the other were discussed. There was hardly any willingness to listen on either side: at the time of the Maastricht norms, the industry considered itself to be no more than the financial money-spinner for the health insurance sector. The government and health funds accused the industry of manipulating the prescription behavior of doctors in order to rake in huge profits at the expense of the health insurance budget.

As always, the truth was and is somewhere in the middle. It is certain that – in a rather important number of cases – this period of non-consultation for both parties has caused irritations, frustrations and for some even indelible marks on the soul, which for the future will remain important in their relations with the other party.

Therefore, in my opinion, it was an action of great merit by former Secretary of Social Affairs Frank Vandenbroucke to definitively abandon this contra productive strategy of confrontation. Technically, this meant accepting for the first time the industry as a full partner in the health insurance consultation model and giving it actual voting authority during the inauguration of the Commission for Reimbursement of Medicines (CRM). I won't conceal from you that the decision by a socialist minister to accept representatives of free market capitalism as full actors in one of the most important decision-making bodies of our solidarity financed health insurance system, was received with quite some surprise and even horror within my organization and also in the other health funds.

Moreover, Frank Vandenbroucke saw to it that short shrift was given to the drafting of completely unrealistic budgets for the pharmaceutical sector we knew in the nineties and which heavily taxed the management of health insurance during those years. That initiative also showed vision and political courage, since his opponents could have easily blamed him for budgetary lapses. For the first time, there was a constructive dialogue between government and industry about the real and realistic needs in the field of medicines and at the same time structural measures were taken to control expenditure in the sector, for example, the system of reference reimbursement. Vandenbroucke's successors have continued this policy with success, with the result that for the last few years we've had a well-balanced budget.

So Frank Vandenbroucke and his successors saw to it that in a first stage, government and industry talked and listened to each other. Now we have reached the stage where they even seem to show some comprehension and empathy for each other's concerns and expectations.

However, we are still far from the most difficult phase, namely that both parties will permanently take into consideration the expectations and concerns of the other party in their respective policy and strategy. However, there seem to be indications that both parties are willing to make a move in that direction and are prepared to become partners instead of parties with conflicting interests.

From an objective point of view, recent evolutions indicate that the time of the 'easy' blockbusters – abstraction made of the lifestyle but barely reimbursed drugs – which are included in the insured package, is a thing of the past for the industry. If technological breakthroughs can be expected, it will more likely happen in the niche markets with specific and well-defined target groups or in the sector of orphan drugs. The investment risks, along with continually more stringent demands of regulators in the field of security and testing, are increasing and the return is very unsure. On the one hand, this is due to the fact that the potential market becomes much more specific and thus smaller, and on the other, because the means of public financiers become relatively limited in volume and therefore it is not at all certain that the government will continue to be willing to accept certain new medicines in the insured package, even if their therapeutic added value is clearly established. For the shareholders it can thus be interesting that long-term arrangements are settled concerning the financing and integration in the reimbursement package of certain innovative medicines that are being or will be developed.

It also becomes easier for the government, given the challenges of the ageing population and the limited financial means disposable, to have a long-term idea about what can be expected in the field of innovation and what is socially relevant to accept for coverage.

Ideally, such agreements should run parallel to the development cycle of a new drug. However, politically such a long-term vision is completely unrealistic: agreements embracing an election term, however, seem realistic to me.

In order to have a chance of success and in order that there would be enough confidence between both parties, these agreements and arrangements should be based on corporate governance principles, which implies the following concrete rights and duties for the government and industry:

- The industry agrees to inform the government of medicines that are expected to come onto the market during an election term.

- Provided that all possible guarantees are given with respect to company secrets, the industry agrees to make the price setting of its medicines transparent. If these arrangements had already be in place, it would possibly not have taken until November 2007 to reimburse Gardasil®. A price setting comparable to that in Australia – nearly half the Belgian price – would probably have resulted in much earlier reimbursement.
- Finally, the industry agrees to include all publications and studies about the therapeutic evaluation of new medicines in the registration file. Today, we often find that the industry limits itself to its own selection of publications – or sometimes even prohibits the publication of articles with a negative evaluation – making it impossible for the government to get an overall idea about the potential therapeutic added value and/or alternatives. In this respect, the Herceptin case still leaves a bitter aftertaste: in spite of the irrefutable breakthrough this medicine represents for the chances of survival of a specific group of patients suffering from breast cancer, the final decision-making would have probably been different if the government had been able to take into account the results of the Finnish nine-week study.
- When scientific or therapeutic arguments *in fine* lead to a price cut of a registered drug, the industry should have the decency not to try to overcome this price cut and the consecutive drop in sales by the well-known policy of so-called 'me-too' drugs. Otherwise, the government and industry will never find the necessary trust base.

When all these conditions are fulfilled – which will not be easy, given the enormously delicate and confidential information it involves divulging for the industry and the important legal adaptations such a procedure requires for the government – the government could agree to:

- A fair return on investment for the innovative medicines that it knows will come onto the market within its term of office.
- A faster reimbursement procedure for these medicines, where every form of 'commissioning' is strictly forbidden for budgetary reasons. This must make it possible to avoid painful situations such as Prevenar®, when the Federal Knowledge Centre was first requested to make an evaluation about the added value of the vaccine before proceeding to the reimbursement, but when after a deadly incident this indispensable evaluation was suddenly no longer so indispensable for reimbursement purposes.

Finally, the government and the industry should sit around the table to agree an honest price setting for patent-expired medicines. From the echoes we receive from the field, which mention important promotion campaigns from the producers and distributors of generic medicines targeting GPs and pharmacists, we are in fact inclined to think that present price levels still allow important margins.

I think that in time such long-term agreements will result in a win-win situation for both parties. Should these mechanisms be successful and create a relationship based on trust between the financing government and the industry, it could also be envisaged (if the political will is there), to organize this system on a supranational, European level. However, that is probably still far ahead in the future.

Conclusions

Allow me to conclude with the following points of interest:

1. The ageing population and the specific financial situation of Belgian health insurance force us to fundamentally reconsider relations between the different actors, that is if we want to keep our health care system qualitative, affordable and accessible for all patients. This is entirely the ambition of my organization. In the past, it has often become clear that in such delicate situations there is sufficient creativity to be able to handle the situation. I trust this will also be the case now. Moreover, since we will be the first to be confronted with the social consequences of the ageing population, we will also be the first to achieve the necessary expertise in this matter and to export it afterwards. The importance of this potential market will only be clear in 20 to 30 years time, when the demographic consequences of, for instance, the Chinese one-child-policy at the end of the seventies will be fully seen.

2. Belgian health insurance has already integrated many elements of corporate governance, even before this expression existed. As with all management concepts that seem to have any success, this is after all an application of simple common sense in your decision-making process. In fact, the pragmatic approach that is typical of the Belgian health care system has introduced this over the years.

3. If we want to succeed in our purpose to keep health insurance affordable and accessible, then the industry should take the plunge with us. The industry has been treated in a step motherly way for years and very often has not been averse to doubtful practices. But the way sectorial evolutions are now looking indicates that it is possible that it can also benefit from such a long-term agreement, one that would finally lead to full partnership.

BIO

Guy Peeters is General Secretary of the NVSM (National Association of Socialist Mutualities) and Chairman of the Board of Directors of P&V-verzekeringen, member of the Board of Directors of the Autonome Hogeschool Antwerpen, member of the high council of the University of Antwerp and President of the Board of Directors of VRT, the public broadcasting network of the Flemish Community.

CO-PAYMENT AND FUTURE SOCIO-FISCAL MODELS

ORVILL ADAMS & VERN HICKS

Virtually all countries have a mix of public and private participation in both the provision and financing of health care. The characteristics of this public-private mix vary as a result of many factors, including the extent of government involvement in health care, which tends to follow a gradient that is closely associated with the strength of government involvement in social and economic systems. In many countries government in the past assumed responsibility for health care provision due to a political philosophy about the role of government (e.g. socialist systems) or a lack of institutional capacity to support private practice. In some developing countries, health care has been centered in government clinics and there has been a lack of economic capacity or social infrastructure to support private practice.

During the last two decades, the distinction between public and private health care providers has been blurring as private sector provision has grown in most countries and governments have sought new models of health care provision and finance. Where public providers depend on private finance (user charges, etc), financial incentives may be a strong influence on the amount and type of care provided. Private providers are often contracted by governments to provide care complementary to traditional public sector care.

Characteristics of public-private mix in health care

McKee et al. (2006) argue that the delivery of health care in almost every country involves some form of public-private partnership. Evolving partnerships were driven in the eighties by *"an emerging neo-liberal consensus that sought to reduce the role of the state"*. The response in the health sector, suggests McKee et al., was to develop quasi-markets and separate purchaser and providers within the public sector. The objective of greater involvement of the private sector was and continues to be increased value for money, improved quality and the delivery of client-friendly services.

One of the most prevalent forms of private finance for health care has been a user fee. The debate about user fees has been more ideological than financial (Hjertquist, 2002). The experience with patient user fees has not been encouraging, especially in terms of access by the poor (Creese et al., 1995). Policy attention is shifting to methods for risk sharing as a means to enable access in the face of user charges or to introduce innovative payment methods (WHO, 2002).

Economic forms of health provision include public sector delivery, for-profit firms and not-for-profit organizations. Risks of inappropriate outcomes in private care are greater according to the form of private ownership and the degree of professionalism in the provision of health services. Public sector oversight, in the form of regulation or contracting, is usually required in the case of for-profit provision of health care. Otherwise, the profit seeking interests of firms may lead to strategies that decrease professionalism and quality of care. Higher risks and costs have been documented in investor owned hospitals and nursing homes where profitability is the key objective (International Herald Tribune, September 25th, 2007).

Categorization of public-private mix

Most financing and provision models in health care involve a mix of public and private sector organizations (Table 1). Public sector delivery is usually associated with public financing, although user fees are common in many countries. Private sector delivery may be financed through either public or private insurance, or a combination of both. Even where there is a national health services model, there is usually also private delivery, often by providers who practice within the public system. Consequently, there are few if any countries that have a system characterized by complete public or private responsibility for health services delivery and finance.

Table 1

	Public (collective payment)	Private (individual payment)	Finance
Public sector delivery	National health service	User fees for public services	Public finance (collective payment)
Private sector delivery	Public Insurance	Private Insurance	Private financing (individual payment)

Forms of ownership follow a gradient from purely public sector ownership to non-profit and profit oriented private ownership (Table 2). Private for-profit includes professional practices, where profit seeking is mediated by professional standards and ethics. Private corporations are closer to the neo-classical model of firms that are primarily motivated by profit seeking and are responsible to shareholders. Social control usually rests with state agencies or boards that are expected to serve the public interest and enforce regulations. These regulations usually give legal sanction to standards of service or seek to protect the public interest in a variety of ways.

Table 2

	Ownership/Control	Social Control
Public	State/national/region/local	State agencies/ Quasi state
Private non-profit	Community	State agencies/ Quasi state boards
Private for-profit	Private one or more/ Shareholders	State agencies
Private small business	Single or partnerships	State agencies
Private corporation	Investor owned/ shareholders (100% owned – not on stock market)	Financial regulatory agencies/state agencies??

Drivers of growth in private involvement

A number of factors have driven growth in private involvement in health finance and provision:

- Escalating health care costs in the public sector have led governments to accept or encourage greater degrees of private provision and finance. Private income growth has been an enabling factor in this trend, as higher incomes have led to an increase in demand for quicker access, high technology services and services with amenity value, such as upgraded facilities in private hospitals.
- Attitudes about management – New Public Management has advocated using private sector management methods in the public system (Moore). This trend has led to a greater emphasis on incentives than on rules, including attempts to create internal markets within public systems. The application of new public management has led to a greater stress on effective management than on forms of ownership. The ideological divide between public and private control is blurred in this approach and resistance to an increased role for the private sector tends to diminish.
- Neo-conservative political and social philosophies have stressed a greater role for markets and competition in the provision of health services.
- Pressures for lower taxes in developed countries have reinforced neo-conservative policies and concerns about rising costs. This set of circumstances has created a climate that enables increases in private responsibility for health care.
- Lack of resources in the public sector has been an important factor in developing countries, many of which have experienced financial crises as both political and economic systems have changed rapidly.

There has also been a perception that the public sector lacks the capacity to adapt to change and to find solutions to the challenges that have often accompany rapid economic change.

Benefits and consequences of public-private mix

Proponents of greater privatization in health care often cite the faster response of the private sector in meeting changing health sector requirements. While the new public management model tends to diminish the strength of this perception, it is also the case that considerable managerial and institutional capacity is required in new public sector management models, whereas private sector actors can often use ingenuity to meet new or existing challenges in health care in an incremental fashion.

Private sector finance provides a source of new money to finance growth in health services. For health care institutions, private finance provides access to equity investments and debt financing in order to expand and modernize facilities. Advocates of private financing also argue that private finance should moderate cost expansion as private sector firms employ analytical techniques such as benefit/cost analysis to develop business cases for new investments.

More efficient deployment and management of resources tends to be associated with private sector management, at least in theory. Characteristics of private sector management also include an increased role for new ideas and managerial techniques, and quicker uptake of new technology – all of which lead to greater innovation and efficiency. It is worth noting, however, that in the economic theory of the firm these characteristics of private sector management require competition and consumer sovereignty. A transition from public sector management to management by a private sector monopoly will not guarantee an increase in innovation and efficiency. Private delivery models may expose a vulnerable public that does not have sufficient expertise to judge the quality of health care to profit seeking behavior that is not in the public interest.

Other consequences and risks of private health care include the potential for a shift in values from solidarity to market forces, which can be problematic in the case of essential services that often require communal financing and subsidized access for the poor. Increasing privatization can also lead to diminished political and financial support for public insurance or subsidies.

There is also a danger that privatization will be seen as the preferred method of dealing with problems in public sector delivery and finance, leading to an abandonment of policies that seek to create efficiencies in the public sector. There are also dangers of trade-offs between quality and efficiency in public sector delivery models that stress the production of services at the lowest cost. An example would be reducing the number of professional nurses and increasing the roles of untrained, lower paid staff in investor owned long-term care facilities. Lastly, a focus on profitable services carries with it a danger of avoiding the sick and the poor, criticisms that have been made of privately owned hospitals.

User fees in health care are a form of private finance that has attracted much controversy. Proponents see user fees as a way to reduce frivolous use and help to make individuals more responsible in their use of health services. User fees are also seen as a way to moderate demand and to mobilize more money for health finance.

Research has led to criticisms of user fees, both for their redistributive effects and their questionable efficiency as a means of increasing financial resources for health care. User fees result in a redistribution of income from the poor to the more affluent (Evans, 1995). They restrict access to needed services for the poor (Nanda, 2002), (Creese et al., 1995). The costs of collecting fees often out-weigh revenue generation objectives (Creese, 1997). User fees shift focus away from population-based risk sharing arrangements to selective individual risks. The introduction of user fees in public systems introduces a new financial dimension in doctor–patient relationships and may discourage seeking types of care that are not perceived as urgent, but are important for long-term maintenance of health. The provider is also placed in a situation of moral hazard where they are able to take advantage of the patient–doctor relationship and realize rents that are not justified or prohibited. This relationship in the face of user fees must be managed through regulations, monitoring, information to the patient and ethical based practice.

The weight of evidence on user fees, especially from developing countries, indicates that user fees for basic health services are a burden on the poorest. Nonetheless, user fees continue to be used as a policy choice by governments in many countries. User fees are growing in public sector programs to finance pharmaceuticals, social support home services, dentistry as well as hospital and physicians care.

Governments fulfill a regulatory and an enabling role in private sector health care and finance. The traditional regulatory role includes acting as a counter force to private interests in order to protect public interests. Such regulation includes enforcing standards of safety and quality as well as preserving access to services by all sectors of the population.

Governments set policy and determine direction for the health sector. In order to fulfill these roles, public bodies require information from the private sector for system planning and monitoring of results. It is now widely recognized that cooperative relationships between governments and the private sector can serve the legitimate interests of both.

The keys to successful collaboration

The recent movement to promote public-private partnerships has grown from the realization that the objectives and responsibilities of both sectors can be furthered by active cooperation in formal partnerships. Formal agreements usually take the form of contracts, although models differ. In the traditional contracting model, one group contracts with another to provide specified services. In a national health system, private sector actors may be contracted to provide care or manage health care institutions under specified terms and on financial terms negotiated between both parties.

Analyses of contracting in health care have found that there is a challenge in constructing and managing contracts for private sector care (Hicks, 2000). These arrangements can be challenging in countries that lack institutional capacity or do not have a cadre of skilled managers in the pubic sector. Success factors in countries that have used contracting successfully include flexibility throughout the term of the partnership agreement. Although contracts are valued in theory for advancing competition, recent experience in the NHS has found that long-term cooperative contracts are superior to competitive contracting in health care. Conditions for successful contracting arrangements include the need to design and balance private provider incentives in line with public goals. Effective and appropriate regulation and legislation are also required in order to ensure that the goals of the contracting body are met while not stifling the contractors' creativity and freedom to act.

Public-private partnerships can also include financing and/or management agreements, for example where private firms finance and manage construction of a health care facility that is then leased to the public sector under long-term agreements. Such agreements may also include provisions to have the private partner manage certain features of the facility that respond well to the use of business principles (e.g. administrative and support functions of a hospital complex).

Other public-private partnerships may have a broad international focus, such as partnerships to extend drug therapy to developing countries. The GAVI Alliance is an example. The Alliance partners include UN agencies (WHO, UNICEF, the World Bank), civil society organizations, public health institutes, donors and implementing country governments, the Bill & Melinda Gates Foundation, and other private philanthropists, and vaccine industry representatives. Successful public-private partnerships can pose challenges to both partners in adapting different value systems to cooperative planning and management of results. This adaptation can require substantial investment of skills and time to achieve success.

Obstacles that must be overcome in public-private partnerships include reconciling different accountability mechanisms for corporations and governments. Both sides have unique mandates that must be respected, including accountability to the public and to shareholders. Trust relations and flexibility of both parties in accommodating the interests are required throughout the partnership agreement. These relations are often difficult to forge and maintain. Evidence shows that development of sound contracts requires skill, diplomacy and measurement systems – this is not easy.

Keys to success in managing public-private partnerships are summarized in Table 3, which was developed from an analysis of partnerships in collecting and managing health information.

Table 3   Keys to success in public-private collaboration

Credibility	Managing hidden agendas
Shared vision	Incremental approach
Critical mass	Demonstrated practical value
Demonstrated commitment	Defined focus (author's interpretation)
Leadership	Responsible public reporting

Source: Milbank Memorial Fund 1999

While both government and the private sector have high level concerns (e.g. policy mandates and maintaining profitability), there are also concerns around the process of public-private partnerships. Private sector partners have identified a number of factors that may impede success or cause reluctance to participate in partnerships (Windmill 2007, Harvey et al). These include a perceived lack of a fair competitive environment if governments dominate the relationship. Different treatment in terms of standard requirements and inspection regimes have also been cited, with a concern that regulatory bodies may feel obliged to require higher standards from partnerships in order to demonstrate that there is no favoritism accorded to activities in which the public sector is involved.

Private sector partners may be frustrated by the fact that the public sector is constrained in its responsiveness to change by the need for lengthy public consultations. Independent providers may withdraw from contracting agreements and potential entrants may be discouraged because of high uncertainty. On the other hand, there is often a perception that the private sector is not trusted to deliver mainstream services and government agencies are constrained from fully exploiting the potential of partnerships in some care sectors as a result.

Some analysts have concluded that there are constraints to developing a productive interface between public and private sectors due to the existence of two silos and a lack of dialogue due to disparities in value systems (e.g. decision-making processes), minimal understanding by each sector of the other sector and mutual distrust based on real or perceived adversarial relationships in the past (Nestman et al).

Opportunities for the future

There are many opportunities for expanding public-private partnerships. One fundamental condition that must be met in the health care sector is the requirement that public sector policies protecting the poor, preventing disease and promoting health must be paramount in any public-private relationship. Once this basic condition is met, developing more innovative models that feature strengthened trust relationships and accountability mechanisms can expand the scope of partnerships.

Other enabling factors in expanding public-private partnerships include:

- Mechanisms for sharing of experiences between the public and private sectors through the exchange of staff.
- Common databases available to the public so that they can make informed choices.
- Common databases across health sub-sectors (dentistry, pharmacy, home care and alternative practitioners) in order to better manage population.

New opportunities for public-private partnership are being recognized in the area of health research, both in terms of funding and carrying out research in high priority areas.

Multidisciplinary non-ideological research into the nature and characteristics of public-private partnerships is required to inform policy decisions in government, boardrooms and civil society organizations.

REFERENCES

Evans R.G, Braer M.L, Stoddart G.L. *"User fees for health care: Why a bad idea keeps coming back (or, What's health got to do with it anyway?)*, Journal of Aging 14(2) 360-390, 1995

Creese A. *"User Fees"*, British Medical journal 1997; 315:2002-2003

Creese A, Kutzin J. *"Lessons from cost recovery in health"*, Geneva: World Health Organization, 1995. SHS/NHP. Forum on Health Sector Reform (Discussion Paper No.2)

Deber R. *"Delivering Health Services Public, Not-for- Profit, or Private"*, Discussion Paper No. 17, Commission on the Future of Health Care in Canada, August 2002

DFID Health Systems Resource Centre and London School of Hygiene & Tropical Medicine; *"Making the most of the private sector"*, August 2000

Gruber J. *"The Role of Consumer Co-payment for Health Care: Lessons from the Rand Health Insurance Experiment and Beyond"*, Kaiser Family Foundation, October 2006

Hicks V. *"Public Policy and Private Providers in Health Care"*, (unpublished), 2001

Hjertquist J. *"User Fees for Health Care in Sweden: A two-tiered threat or a tool for solidarity"*, Atlantic Institute for Market Studies, Health Commentary No. 6. 2002

International Herald Tribune, September 25th, 2007

Lonnroth K. et al. *"Public-Private Mix for DOTS Implementation: What makes it work?"*, Bulletin of the World Health Organization, August 2004, 82(8)

McKee M, Edwards N, Atun R, *"Public-Private Partnerships for Hospitals"*, Bulletin of the World Health Organization, November 2006, 84(11)

Moore M. *"Managing for value-organizational strategy for profit, non-profit and other governmental organizations"*, Non-Profit and Voluntary Sector Quarterly, Vol 29 No1. pp. 183-204

Nanda P. *"Reproductive Health Matters"*, Efsevier Science Ltd. 2002;10(20) 127-134

Nestman L and Joffre C. *"The Medicare and Private Baskets of Health Services in Canada: two Silos and No Dialogue"*, 9th International Conference on Priorities in Health Care, Wellington, New Zealand. November 2004

Tuohy C, Flood C, Stabile M. *"How does private financing affect public health care systems? Marshalling the evidence from OECD Nations"*, www.chass.utoronto.ca/cepa/priv..2007

Milbank Memorial Fund Reforming States Group *"Public-Private Collaboration in Health Information Policy"*, June 1999, www.milbank.org/reports/mrinfopolicy.html

World Health Report, 2000. Overview. World Health Organization. Geneva

BIOS

Orvill Adams is the Director of Orvill Adams & Associates, a consultancy based in Amsterdam, the Netherlands. He brings over 20 years of experience in senior positions in the health sector and has spent 10 years with the Canadian Medical Association as Director of the Department of Medical Economics. Adams was with the World Health Organization for 10 years as Director of the Department of Health Services Provision. He has managed and developed health systems, workforce policies and tools, and has interacted with Development Partners, bilateral and multilateral, at global level and within countries. Orvill Adams is widely published in the area of health workforce development, management, policy and planning and holds an adjunct teaching position at the University of Helsinki, Department of General Practice. He holds postgraduate degrees in Economics and International Affairs.

Vern Hicks is a consultant in health economics, health information and health workforce planning. He has many years experience in international work on projects for the World Health Organization, the World Bank and the Canadian International Development Agency. He has an adjunct faculty appointment at the Department of Community Health and Epidemiology at Dalhousie University, Halifax, Canada.

WHY WE NEED A NEW APPROACH TO PHARMACEUTICAL INNOVATION: A PRAGMATIC ANSWER TO A MORAL QUESTION

THOMAS POGGE & DORIS SCHROEDER[1]

Nobel Laureate Milton Friedman famously declared: "The social responsibility of business is to increase its profits"[2]*. He denied that company leaders had any social responsibility such as eliminating discrimination, avoiding pollution or creating jobs. In his view, those who advanced such ends alongside the generation of profits were "unwitting puppets of the intellectual forces" of socialism*[3].

When the pharmaceutical industry is attacked for making *"profits that kill"*[4] or is publicly indicted for its alleged complicity in the African public health crisis as in the film 'The Constant Gardener', Friedman might shrug his shoulders and repeat the mantra: *"The social responsibility of business is to increase its profits"*. In this chapter, we shall argue that even in Friedmanesque territory, a new approach to pharmaceutical innovation can be justified persuasively.

This chapter is in three parts. Part 1 will outline briefly why, thirty years after Friedman's pronouncement, management experts believe that value-led business will be more successful than business based on sheer profit maximization. Part 2 will describe how the pharmaceutical industry currently fails at least parts of the value-led business test in the estimation of

---

1 The research leading to these results has received funding from the European Community's Seventh Framework Programme under grant agreement 217665. The authors wish to thank Matt Peterson, Armin Schmidt, Julie Lucas and Cathy Lennon.
2 Friedman, M., (1995). *Legitimacy and Responsibility*, in: Hoffmann, W. / Frederick, R. (eds), *Business Ethics – Reading and Cases in Corporate Morality*, 3rd edition, McGraw Hill, New York, pp.137-141.
3 Ibid.
4 Bunting, M., (2001). *Profits that Kill*, The Guardian, online: www.guardian.co.uk/world/2001/feb/12/wto.aids

the general public. Part 3 will introduce a market-based solution, which can align shareholder interests with moral imperatives.

Brutal profit maximization pays only in the short run

In the early 1970s, management guru Peter Drucker[5] famously claimed that the purpose of organizations *"is to enable ordinary human beings to do extraordinary things"*. Ever since, there has been much talk in management circles about combining business with moral imperatives and evidence increasingly shows that it pays. A business strategy driven by ethical values will produce increased profitability. According to Philip Holden[6], employees will only give their best if they can identify with something worthwhile at work and if they are proud of the purpose and the belief-system within their company. According to Blanchard and Peale[7], pride in the strategic vision of one's organization will bring out the best performance in employees and lead to high quality work and profitability. When people have negative feelings about their organization, they often try to *"even things out"* by calling in sick, making private long-distance phone calls, etc.[8] On the other hand, if they are proud of their organization, they will try to maintain its integrity[9]. The key to successful businesses is therefore leaders whose integrity, vision and ethical forethought is strong enough to inspire staff[10]. Ethical business is a means to corporate success, i.e. increased profits.

Yet, employees are not the only ones to whom a business must appeal in order to succeed. Consumers are just as important. Ethical consumers are buyers with at least two preference strategies. First, like all consumers, they show the preference to increase material utility, i.e. to receive a product or service for their money rather than give it to charity. Second, they want to benefit a third party or avoid harming it in the process.

Ethical consumerism is on the rise in the West with six per cent of the UK adult population (2,8 million people) being committed consumers of ethi-

---

5   Drucker, Peter, (2000). cf. Holden, Philip; *Ethics for Managers*, Gower, Aldershot, p.143.
6   Holden, Philip, (2000). *Ethics for Managers*, Gower, Aldershot, p.131.
7   Blanchard, Kenneth and Peale, Norman V., (1990). *The Power of Ethical Management – You Don't Have to Cheat to Win*, Cedar Books, London, p.95.
8   Ibid. 84.
9   Ibid. 95
10 Holden, p.143.

cal products and services[11]. This is the case in instances as diverse as animal testing for cosmetics, tropical timber sales, ethical banking or free-range eggs, and even pharmaceutical products. The UK's leading alternative consumer organization (Ethical Consumer) advises consumers who want to take pharmaceutical companies' records on social and environmental issues into account on which non-prescription drugs they ought to avoid. Some of the criteria used are animal testing, excessive profits, research on essential drugs for developing countries, and record on corporate lobbying[12]. This means that ethical consumerism can impact on pharmaceutical profits even though demand curves for pharmaceutical products are generally considered highly inelastic. Companies that ignore those who *"shop with a conscience"*[13] should anticipate a drop in profits as market shares for ethical companies keep rising.

How does the pharmaceutical industry fare on value-driven visions for employees and ethical consumers?

In business, a good reputation matters. The pharmaceutical industry, in general, does not have one. A study undertaken in the United States showed that of media references to the industry, 57% were unfavorable, 18% neutral and only 25% positive[14]. Given that the majority of US citizens enjoy the benefits of drugs and services tailored to their health needs, this is a stunning result. One can only imagine what a similar survey would indicate if undertaken in, for instance, South Africa, where the industry undertook a *"disastrous public relations move"*[15] trying to stop the government from procuring cheap drugs to fight their HIV/AIDS crisis. And another study has shown that the pharmaceutical industry is increasingly seen as taking unfair ad-

[199]

---

11 Carslaw, Nicola, (2000). *The Consumer Strikes Back*, BBC News Online: http://news.bbc.co.uk/1/hi/business/959936.stm. Cowe, Roger, (2000). *Morality is a Spending Force*, Guardian Unlimited Archive. Online: www.guardianunlimited.co.uk/Archive/Article/0,4273,4072666,00.html. Co-operative Bank Ltd., *The Ethical Consumerism Report 2007*, p.6. Online: www.co-operativebank.co.uk/images/pdf/ethical_consumer_report_2007.pdf .
12 Harrison, Rob, (2006). *Reflections on reaching 100*, online: www.ethicalconsumer.org/CommentAnalysis/Features/100IssuesofEC.aspx. Also www.ethicalconsumer.org/FreeBuyersGuides/healthbeauty/Coldremedies.aspx.
13 BBC News Online, (1999). *How Can You Shop With a Conscience?*, online: http://news.bbc.co.uk/1/hi/uk/465649.stm,
14 Melé, Domènec (2005) *Do Drug Companies Deserve Their Bad Reputation?*, online: www.bettermanagement.com/library/library.aspx?l=14220
15 Johnston, J., and Wasunna, A.A., (2007). *Patents, Biomedical Research, and Treatments: Examining concerns, canvassing solutions*, Hastings Center Report, Vol.37, No.2, S1-S36, p.S16.

vantage of consumers through excessive drug pricing[16]. In this chapter, we shall look at two main ethical concerns: (a) high drug prices and their impact on global health and (b) potential exploitation of research subjects in developing countries.

"Profits that Kill"

Subsisting on incomes around $100 or $200 per person per year, the poorer half of humankind is highly exposed to life-threatening deprivations. According to official statistics, there are roughly 6,700 million human beings alive today[17]. Of these, some 923[18] million are chronically undernourished, 884 million lack access to clean water,[19] and about 2,000 million lack access to essential medicines[20]. People living with such severe deprivations are bound to be susceptible and vulnerable to infectious diseases and often unable to overcome them. Today, one third of all human deaths are poverty-related: including over nine million children under the age of five[21]. How does this relate to the pharmaceutical industry? Madeleine Bunting writes in *"Profits that Kill"* in The Guardian[22] that:

"Put baldly, patents are killing people. ...Intellectual property protection has become a tool to make permanent the growing inequality of the global economy: the rich get richer and the poor get poorer. Drugs are only the most blatant example of how, through TRIPS[23], the developed countries have stacked the odds in their favour."

[200]

---

16 *Weiss Ratings Reports High Level of Blame Against Pharmaceutical Companies For Excessive Drug Costs*, Insurance Advocate 2/23/2004, Vol. 115 Issue 7, p22.
17 Online: www.census.gov/ipc/www/popclockworld.html
18 UN Food and Agriculture Organization, (2008). *Briefing Paper: Hunger on the Rise*, online: www.fao.org/newsroom/common/ecg/1000923/en/hungerfigs.pdf
19 UNDP, (2007) *Human Development Report 2007/2008*, Houndsmills: Palgrave Macmillan, p.254,. www.wateraid.org/uk/what_we_do/statistics/default.asp
20 Hollis, Aidan, and Pogge, Thomas, (2008). *The Health Impact Fund: Making New Medicines Accessible for All*, Incentives for Global Health, p.114
21 Karwal, Roshni, (2008). *Policy advocacy and partnerships for children's rights*, online: www.unicef.org/policyanalysis/index_45740.html
22 Bunting, M., (2001). *Profits that Kill*, The Guardian, online: www.guardian.co.uk/world/2001/feb/12/wto.aids.
23 TRIPS was negotiated in the 1986-94 Uruguay Round, and it took effect on 1 January 1995. Essentially, it demands that common types of intellectual property are recognized and effectively protected through national law, whether the intellectual property right holder is a native or a foreigner within the country. Thus patents, trade marks, designs, copyrights, plant breeder rights, geographic indications, trade secrets and circuit layout rights are protected.

One of the beauties of the human intellect is that its creative products are non-rivalrous, as economists say: the intellectual effort of writing a novel is exactly the same, whether people all over the world or two lone individuals (one's parents) read it. The same applies to music, software, new plant breeds or a newly discovered molecule. Millions can benefit from intellectual efforts without adding significantly to their cost: the production of multiple copies of books, CDs, seeds, pills, vaccines etc. are typically not expensive.

Of course many such efforts would not be possible without the promise of a return to cover the substantial development costs of the product and in order to protect the intellectual efforts of inventors. Therefore most states adhere to the idea of intellectual property that is entitled to legal protection through tools such as copyrights, trademarks and patents. Without such protection, it is assumed that the products would not come into existence.

Patents are therefore seen as *incentives* for innovation. They entitle the patent holder to stop (unauthorized) use of an innovation disclosed in the patent by anyone else, typically for a period of 20 years. In pharmaceuticals, patents are particularly important, since competition with generic products tends to be fierce and the cost of product research and development (R&D) is very large relative to the subsequent cost of production. In a free market system without patents, pharmaceutical firms would be unlikely to recover their R&D outlays and would therefore be unwilling to invest. Medical progress would stall and important new medicines might never be invented. The patent system can therefore be justified on grounds of social utility. Human beings tend to become ill, the patent system procures cures and treatments for such illnesses, and hence patents are acceptable, even desirable. Yet, the patent system does not help all human beings. Given the high monopoly prices charged under patent protection, affluence or citizenship in an affluent country are usually required to access a comprehensive range of medicines.

Since patents can only be justified by an appeal to social utility (it does not make sense to argue that patents are God-given or that inventors have a human right to a 20-year privilege to charge monopoly prices), one needs to ask whether the current regime is still justifiable on the simple premise that pharmaceutical companies have high R&D costs. In other words, how does the global intellectual property rights (IPR) regime, as institutionalized through TRIPS, affect the social utility of diverse human populations? In examining this question, it is crucial to avoid the false dichotomy that asks us either to accept the current global IPR regime or else to renounce all hope for pharmaceutical innovation. For example, a third possibility existed in the recent past, when IPRs were legally recognized in most affluent countries but not in most of the poorer ones.

Before 2005, Indian law only allowed patents on processes, not on products. As a result, India had a thriving generic pharmaceuticals industry that supplied copies of patented medicines cheaply throughout the world's poor regions. However, in 1994 India signed up to TRIPS as negotiated in the Uruguay round of the General Agreement on Tariffs and Trade (GATT) treaty. As a result, India was required to introduce patents on products by January 2005. This change to Indian patent rules hits the world's poor in two ways, directly by undercutting the supply of affordable medicines and indirectly by removing the generic competition that reduces the cost of brand-name medicines [24].

The existence of this pre-TRIPS environment has two implications. First, the social utility argument for the current global IPR regime cannot succeed by showing merely that this regime is preferable to the complete absence of IPRs anywhere. Before 2005, the pharmaceutical industry was reliably supplying drugs without a global IP regime. Second, the social utility argument for the ongoing IPR initiative fails if the decline in social utility it brings for poor populations (by reducing access to essential medicines) is greater than the increase in social utility it brings to rich populations (by maximizing profits). On any plausible conception of social utility, which gives equal weight to the wellbeing of rich and poor human beings alike, the new global IPR regime is greatly inferior to its more differentiated predecessor.

But if the new regime is so much worse for the global poor, why did they agree to it? To understand this, one must bear two points in mind. First, in the negotiations that preceded the WTO TRIPS Agreement, the representatives of the poor countries were *"hobbled by a lack of know-how. Many had little understanding of what they signed up to."*[25] Poor-country representatives were

---

24 Editorial. (2005 January 18). *India's Choice*. The New York Times..
25 *White Man's Shame*. (1999 September 25). The Economist., p. 89.

facing some 28,000 pages of treaty text drafted in exclusive ('Green Room') consultations among the most powerful countries and trading blocks.

The second point is that developing countries are heavily stratified. Even if an international treaty is disastrous for a country's poor, signing up to this treaty may nonetheless be advantageous for this country's political and economic elite. They may gain diplomatic recognition and political support, buy more arms, protect their ability to transfer wealth abroad, etc. Consent by the ruling elite is not then a valid indicator of advantage to the general population.

On any plausible conception of social utility, the rich countries' IPR initiative goes in the wrong direction, causing many foreseeable additional premature deaths among the global poor by cutting them off from life-saving patented medicines. 'Profits that kill' is perhaps a drastic way of putting it, but the fact that 2,000 million human beings have no access to essential medicines means that (a) the current IPR system is morally unacceptable, given that there was a recent alternative, which was superior in terms of global social utility, and (b) that even the most cynical Friedman follower would have to note that the resulting public image problem of the pharmaceutical industry cannot be good for business, neither in the sense of getting the best from staff nor in terms of attractiveness for ethical consumers. In the final section, we shall introduce an alternative to the current IP regime, which is compatible with TRIPS, yet wins easily on global social utility, even compared to the pre-TRIPS regime. First, though, we shall stay in our Friedmanesque territory and outline one more area for serious reputational concerns.

Guinea pigs for the rich

John Le Carré's book and the later film 'The Constant Gardener' charged the pharmaceutical industry not only with complicity in the African public health crisis but also with the exploitation of naïve and cheap research participants. In an article entitled *"A lot of very greedy people"*, he wrote[26]:

---

26 Le Carré, John. (2001 Feb 12) *A lot of very greedy people*, The Guardian. Online: www.guardian.co.uk/world/2001/feb/12/aids.wto.

> "I had not been exploring Big Pharma for more than a couple of days before I was hearing of the frantic recruitment of third world 'volunteers' as cheap guinea pigs. Their role, though they may not ever know this, is to test drugs, not yet approved for testing in the US, which they themselves will never be able to afford even if the tests turn out reasonably safe."

Is this a novelist's wild imagination? No. On the one hand, highly unethical trials are being conducted in developing countries to benefit those in more affluent countries (see example below). On the other hand, even trials that are conducted according to strict industry guidelines leave behind a sense of exploitation because upon market entry, the products that are being tested are unlikely to be affordable to the study subjects. Let us give two brief examples.

In 1996, an epidemic of meningitis was raging in a poor part of Nigeria, where medication was not readily available. Children were seriously ill and at risk of dying, parents were desperate. A foreign company decided to test a drug, which had not yet been approved domestically, on the children in question. As a result of the trial, eleven children who had taken the new drug died and another 200 became deaf, blind, or lame. Once details of the study became known, the researchers could not produce any documentation to show that informed consent had been obtained from the parents of the enrolled children. Likewise, it became obvious that the trial was "*in apparent violation of established industry guidelines*" (e.g. the administration of the drug and its follow-up were not undertaken appropriately)[27].

Although it was maintained that the death rate would have been similar in a domestic trial, trials like this one should not take place. It is essential that patients or their proxies understand the purpose of a trial and freely consent to take part. If not, the trial is a human rights violation. The legally binding International Covenant on Civil and Political Rights (ICCPR) has been ratified by 152 countries and its Article 7 specifies that[28]:

---

27 Macklin, R. (2004). *Double Standards in Medical Research in Developing Countries*. Cambridge University Press, p.99f.
28 Office of the High Commissioner for Human Rights. *International Covenant on Civil and Political Rights – Status of Ratifications*. 2004. Available online: www.unhchr.ch/html/menu3/b/a_ccpr.htm,.

"[n]o one shall be subjected to torture or to cruel, inhuman or degrading treatment or punishment. In particular, no one shall be subjected without his free consent to medical or scientific experimentation."

In addition, once involved in a study the participants can expect physicians to promote and safeguard their health[29]. If follow-up tests are usually undertaken and required according to industry guidelines, this should be done whether the trial takes place in Nigeria or the United States.

Cases like this one incense readers and greatly contribute to the negative image of the pharmaceutical industry in the general population. Of course, the underlying problem that leads to such unethical trials is lack of access to health care. If Nigerian children had access to essential medicines, companies would be unable to fly into areas of raging epidemics with offers of 'treatment' that is otherwise unavailable.

In this regard, resolving the problem of access to medicines for the poor will also contribute indirectly to the avoidance of unethical trials (for more on a potential solution, see last section). Another problem of potential exploitation is a more common occurrence than unethical trials of the above variety, and has been expressed as follows by one Kenyan research participant who complained to a group of researchers:

"I have been used like a guinea pig, so how does he just leave me without compensation?"[30]

When research is carried out in developing countries by Western pharmaceutical companies, concerns about the unfair distribution of benefits usually arises in two areas. First, the research agenda is driven by the company carrying out the research. This means that decisions about the topic of the research, its location and the recruitment of participants are being taken by foreign citizens from rich nations. The involvement of local expertise is limited. Second, the risks and burdens of the research "are borne by develop-

---

29 Art. 2, *Declaration of Helsinki*. www.wma.net/e/policy/b3.htm.
30 Shaffer, D.N., Yebei, V.N., Ballidawa, J.B., Sidle, J.E., Greene, J.Y., Melsin, E.M., Kimaizo, S.J.N., Tierney, W.M. (2006). *Equitable treatment for HIV/AIDS clinical trial participants: a focus group study of patients, clinical researchers, and administrators in western Kenya.* Journal of Medical Ethics,32:55-60,p.55.

ing countries whose poorest inhabitants serve as research participants, but these countries rarely share the benefits, because many interventions are beyond the economic reach of both the research participants and their governments".[31]

In contrast to the unethical trial described above, the Kenyan cited has probably consented to take part in a research study after receiving full information. Still, any product eventually derived from the research would be far beyond his reach. He therefore feels he has contributed to the advancement of science and the potential development of commercial products without adequate compensation. And he is not alone in this perception[32].

[206] "There is an increasing consensus that, in principle, participants in developing countries should continue to receive benefits originating from the studies in which they enlisted beyond the research period. The consensus is mainly based on... concerns about exploitation and fair benefits... since participants have contributed to others and society, they should be rewarded."

Researchers who do not make the fruits of their studies available to research participants are exposing poor and ill educated people to risks in order to benefit more affluent populations. They are taking advantage of one population to serve another. It is in this context that post-trial obligations have been promoted in order to restore the balance of justice.

In essence, post-trial obligations describe a duty by researchers and research sponsors to provide a successfully tested drug to the research participants after the conclusion of the research study. Mandatory post-trial provision of successfully tested drugs and services to research participants is demanded by the World Medical Association[33], the Joint United Nations Programme on HIV/AIDS[34] and the European group on ethics in science and new technolo-

---

31 National Bioethics Advisory Commission (2001). *When Research is Concluded – Access to the Benefits of Research by Participants, Communities and Countries; Ethical and Policy Issues in International Research: Clinical Trials in Developing Countries*, pp.55-75, p.59. Online: www.bioethics.gov/reports/past_commissions/index.html
32 Zong, Zhiyong (2008). *Should post-trial provision of beneficial experimental interventions be mandatory in developing countries?* Journal of Medical Ethics, 34:188-92,p.188.
33 *Art.14 Declaration of Helsinki.* Online: www.wma.net/e/policy/b3.htm .
34 Joint United Nations Programme on HIV/AIDS. (2000). *Ethical Considerations on HIV Preventive Vaccine Research, Guidance point 10, "Benefits"*, p.30. Online: http://data.unaids.org/publications/IRC-pub01/JC072-EthicalCons_en.pdf

gies to the European Commission[35]. In addition, it is implemented in ethics guidelines in Brazil, India, South Africa and Uganda[36].

Post-trial obligations are one way of addressing potential injustice issues in developing countries and making some drugs available to some of the poor. But are they the best way? We think not. The main practical obstacle to the usefulness of post-trial access is the long time span of pharmaceutical research. By the time a drug enters the market, it might be too late for many research participants to benefit. One also needs to consider the potential for undue inducement if post-trial access to successful drugs is promised to research participants[37]. Those with no or little access to health care are already under pressure to enter trials in order to receive general health care. Adding another considerable benefit could worsen the chances of participants for giving legitimate, genuine consent. It has also been argued that post-trial obligations would make trials in developing countries prohibitively expensive and therefore reduce their numbers by default[38,39]. In a similar vein, it has been argued that post-trial obligations are addressed at resolving issues of the global economy, which is not the task of individual researchers or sponsors[40,41].

---

35 European group on ethics in science and new technologies to the European Commission. (2003). *Ethical aspects of clinical research in developing countries, Article 2.13 "Supply of Treatment After The End Of The Trial"*. Online: http://ec.europa.eu/european_group_ethics/docs/avis17_en.pdf
36 National Health Council of Brazil. (1996). *Resolution No. 196/96 on Research Involving Human Subjects, Article 13*. Indian Council of Medical Research. (2000). *Ethical Guidelines for Biomedical Research on Human Subjects, Article 8*. Clinical Trials Working Group of the South African Department of Health. (2000). *Guidelines for Good Practice in the Conduct of Clinical Trials in Human Participants in South Africa, Article 9*. National Consensus Conference on Bioethics and Health Research in Uganda. (1997). *Guidelines for the Conduct of its Research Involving Human Subjects in Uganda*, Article 12.
37 Participants in the 2001 Conference on Ethical Aspects of Research in Developing Countries. (2004). *From Reasonable Availability to Fair Benefits*. Hastings Center Report. 34(3)2-11, p.6.
38 Brody B., (2002). *Ethical Issues in Clinical Trials in Developing Countries*. Statistics in Medicine.;21:2853-8, p.2857.
39 McMillan, J.R., Conlon, C., (2004). *The Ethics of Research Related to Health Care in Developing Countries*. Journal of Medical Ethics.30:204-6, p.206.
40 Emanuel, E.J., Wendler, D., Killen, J., Grady, C. (2004). *What Makes Clinical Research in Developing Countries Ethical? The Benchmarks of Ethical Research*. Journal of Infectious Diseases..189:930-7, p.935.
41 Ashcroft, R., (2002). *Commentary: Biomedical Research, Trade Policy and International Health – Beyond Medical Ethics*. Social Science and Medicine. 54:1143-4, p.1144.

Instituting post-trial obligations is an attempt at addressing the two major ethical concerns raised, but only when they occur in conjunction ('profits that kill' and the 'guinea pigs for the rich'); in other words, when monopoly pricing restricts access to medicines for those who contributed to a patented product's successful market entrance. In such cases, we have products developed for affluent markets being tested on the poor[42] and subsequently priced out of their reach. This scenario represents the tip of the ethical iceberg. Combating this with post-trial obligations looks like a good start. But, figuratively speaking, post-trial obligations will only provide the symptomatic relief a few idealistic people with weak blow torches can provide. The foundation of the iceberg remains completely untouched. Millions will still die due to lack of access to essential medicines, even though – importantly – the question of direct exploitation of individual human guinea pigs might be addressed. Wanting to tackle the entire iceberg though, we need a much more comprehensive solution than post-trial obligations.

The Reform Plan

The Health Impact Fund would be a new way of stimulating research and development of life-saving pharmaceuticals. To provide wide access to the most effective pharmaceuticals, prices need to be low enough for people to afford – but low prices don't create strong incentives for innovators to invest in research and development. The proposed Health Impact Fund is an optional mechanism that offers pharmaceutical innovators a supplementary reward based on the health impact of their products, if they agree to sell those products at designated low prices. The following provides only a brief sketch of the Health Impact Fund. We encourage readers who would like to know more to visit the project's website, www.healthimpactfund.org.

How would it work?
Pharmaceutical innovators holding valid patents can elect to sell their product globally at a low price agreed with the Fund. In exchange, they will be paid by the Fund annually for ten years based on their product's assessed health impact. Participating firms will also offer zero-priced relevant licenses following the ten years.

---

42 For more on testing drugs on the poor, see Pogge, Thomas. T*esting Our Drugs on the Poor Abroad* in Hawkins, Jennifer and Emanuel, Ezekiel, eds. (2008). *Exploitation and Developing Countries: The Ethics of Clinical Research.* Princeton University Press.

How much would each firm earn?
The low price will be set to cover manufacturing costs, so firms' profits will derive solely from payments from the Fund. Each year, the Fund will have a fixed payout – $6 billion to begin with – to be distributed among the products firms elect to register. This annual payout will be shared among firms in proportion to the assessed global health impact of their drugs in the preceding year. Thus, products will be rewarded strictly in proportion to their health benefits (pay for performance).

What drugs would be included?
The Health Impact Fund would be most attractive for products that are expected to have a large global health impact but relatively low profitability under monopoly pricing. For example, a drug treating a disease mainly afflicting poor people will be an excellent candidate for registration, since typically such products cannot earn high profits, though they could benefit many people. Thus, the Fund will provide important additional incentives to develop drugs for neglected diseases.

How would it affect consumers?
Consumers will benefit from the availability of new drugs at low prices: through reduced cost for national health systems, reduced insurance premiums, and reduced prices at the pharmacy. They will benefit further from increased medical knowledge and better protection against invasive diseases, from the increased concern of pharmaceutical companies with the health impact of their products rather than merely with sales, from reduced counterfeiting incentives, and from massive reductions in the global burden of disease with associated gains in economic productivity worldwide.

How would health impact be assessed?
Health impact can be assessed in terms of a variant of quality-adjusted life years (QALYs), a metric that has been extensively used for more than a decade. Taking the preceding state of the art as a benchmark, the assessment would estimate to what extent the new drug has improved public health worldwide by improving the health of patients who otherwise would have consumed an inferior medicine or none at all. The estimate would be based on data from clinical trials, including pragmatic trials in real-life settings, on tracking randomly selected packages to their end users, and on statistical analysis of sales data as correlated with global burden of disease data.

These estimates would necessarily be rough, at least in the beginning. But so long as any errors are random, or at least not exploitable by registrants, the incentives provided by the Fund would be only minimally disturbed.

Would patent rights be affected?

No. Innovators retain their patent rights. They can elect to give up the freedom to charge monopoly prices in exchange for Health Impact payments from the Fund. Firms will probably make this choice only when they expect higher profits from these payments than from monopoly prices.

How would the Fund be financed?

Governments and other donors would commit to long-term funding. Savings on medicines that would otherwise have been bought at much higher prices will offset some of the Fund's cost to taxpayers. But its greatest benefit is that patients will gain access to important medicines that, without the Fund, would have been too expensive for them, or even non-existent.

Conclusion

The pharmaceutical industry does not have a good reputation amongst the general public. Studies have shown that pharmaceutical companies are seen to take unfair advantage of consumers through excessive drug pricing and that only 25% of media references to the industry are positive. The claims that they generate 'profits that kill' and recruit 'guinea pigs for the rich' are partly responsible for this reputational crisis. In this chapter, we showed that even in Milton Friedman's mantra that *"the social responsibility of business is to increase its profits"*, urgent action is required. Companies will get the best out of their employees and satisfy increasingly ethically sensitive consumers if they display integrity, vision and ethical forethought.

Within the framework set by Friedman, it does not make sense to blame companies for wanting to increase their own profits. Instead one needs a solution that aligns shareholder interests with moral imperatives. The only existing solution without medium- or long-term disadvantages[43] is the Health Impact Fund. Financed by governments, the Fund would offer patentees the

---

43 Nathan, Carl, (2007). *Aligning Pharmaceutical Innovation with Medical Need*, Nature Medicine, 13(3) pp. 304-8.

option to forgo monopoly pricing in exchange for a reward based on the global health impact of their new medicine. By registering a patented medicine with the Fund, a firm would agree to sell it globally at the lowest feasible price of production and distribution. In exchange, the firm would receive, for a fixed time, payments based on the product's assessed global health impact. The arrangement would be optional and it wouldn't diminish patent rights, it therefore aligns the interests of pharmaceutical companies with the interests of poor patients, removing the 'profits that kill' reproach directly and the 'guinea pigs for the rich' indirectly, as research participation would no longer be the only way to access drugs for many. Social justice shakes hands with Milton Friedman. Surely this has to be a winner!

BIOS

Having received his PhD in philosophy from Harvard, **Thomas Pogge** writes and teaches on moral and political philosophy and Kant. He is Leitner Professor of Philosophy and International Affairs at Yale University, Research Director at the Oslo University Centre for the Study of Mind in Nature (CSMN), professional research fellow at the Centre for Applied Philosophy and Public Ethics (CAPPE), and Adjunct Professor of Political Philosophy at the University of Central Lancashire, UK. His recent publications include 'Politics as Usual,' Polity 2009; 'World Poverty and Human Rights', second edition, Polity 2008; 'John Rawls: His Life and Theory of Justice', Oxford 2007; 'Freedom from Poverty as a Human Right', edited, Oxford 2007; and 'A Companion to Contemporary Political Philosophy', co-edited, Blackwell 2007. His work has been supported by the John D. and Catherine T. MacArthur Foundation, the Princeton Institute for Advanced Study, All Souls College (Oxford), and the National Institutes of Health (Bethesda). He is editor for social and political philosophy for the Stanford Encyclopedia of Philosophy and a member of the Norwegian Academy of Science. With support from the Australian Research Council, the UK-based BUPA Foundation and the European Commission (7th Framework Programme), he currently heads a team effort toward developing the Health Impact Fund, a complement to the pharmaceutical patent regime that would improve access to advanced medicines for poor people worldwide, see www.healthimpactfund.org.

Doris Schroeder was educated in Germany and the United Kingdom, in economics, management, philosophy and politics. After working as a strategic planner for Time Warner, she currently holds two professorial appointments. She is Professor of Moral Philosophy and Director of the Centre for Professional Ethics at the University of Central Lancashire, UK and Professorial Fellow at the Centre for Applied Philosophy and Public Ethics at the University of Melbourne, Australia. Her research interests include international justice, human rights and health and benefit sharing. She is currently leading an international EU-funded project on benefit sharing for human biological resources. Doris is a member of the British Food Ethics Council, the Human Rights Editor of the Cambridge Quarterly of Healthcare Ethics, an International Steering Group Member of the Dutch Research Council, a regular ethics review panel chair and expert evaluator for the European Commission, a founding member of the Society of Applied European Thought and a member of the British Philosophical Association (BPA) and the Association of Legal and Social Philosophers (ALSP). For her publications, see www.uclan.ac.uk/cpe.

CHALLENGES AND OPPORTUNITIES OF CANCER CLINICAL RESEARCH AT PAN-EUROPEAN LEVEL

FRANÇOISE MEUNIER

The European Organisation for Research and Treatment of Cancer (EORTC) is a pan-European, non-profit, independent research organization that develops, conducts, coordinates and stimulates high quality translational and clinical research aimed at improving the standards of treatment for cancer patients in Europe. Driving independent pan-European cancer clinical trials poses multiple challenges and opportunities which are presented in this article.

The current picture: Cancer in Europe

Despite of tremendous progress made in the last two decades, cancer remains a major threat in Europe as this is mainly a disease of ageing. Taking into account epidemiological data, it is indeed expected that within EU countries the incidence of cancer diagnosis will increase.

Currently, there are more than 3,191,600 European citizens diagnosed with cancer and 1,703,000 persons die as a result of cancer on a yearly basis.

This increased incidence is mainly related to the ageing of the population but also to the smoking patterns of women, as well as to improved diagnostic tools that have lead to the detection of a significant number of cancers earlier than in the past.

Continued emphasis on prevention, early detection and improved treatment strategies has already resulted in much better cancer control and a major improvement of survival rates as compared to the survival rate in 1970. Cancer must now be considered as becoming the most common chronic disease at least in EU.

Figure 1
Estimated incidence of most common cancer types in the EU-25 in 2006

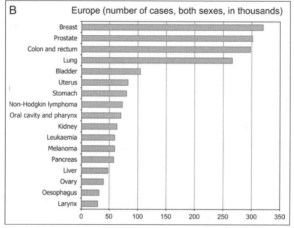

No data for Europe for all the individual sites due to limitations of coding scheme employed.
Source: Ferlay J et al., Annals of Oncology 2007

Figure 2
Five years survival (%)

EORTC 5 YEARS SURVIVAL (%)	1970	2008
• Leukaemia in children	0	80
• Leukaemia in adults	0	45
• Bone cancer	5	60
• Advanced testis cancer	0	95
• Breast cancer	40	85
• Colon cancer	30	60
• Hodgkin's disease	10	90
• Childhood cancer	61	80
• Pancreas	5	10
• Non-small cell lung cancer	0	20

The need for high quality, international clinical and translational research

Emphasis on quality of life and long-term survival is of paramount importance but there will be no further progress unless comprehensive research is conducted in Europe.

Basic research is essential in order to understand the mechanism of oncogenesis and to develop more effective drugs. However, it is an illusion to believe that basic research is the solution to this major health problem. Basic research should be complemented with translational research and high quality clinical research to change practice and thereby outcome.

Taking into account the major breakthroughs in molecular biology, it is expected that targeted medicine will be extremely relevant in establishing more effective treatments for patients with cancer.

Funding for translational research and clinical research is commonly more difficult to find than for basic research, while competition between basic research and clinical research is clearly detrimental as the two disciplines are complementary to define new standards of care.

Figure 3
Medical practice and medical research depend on each other

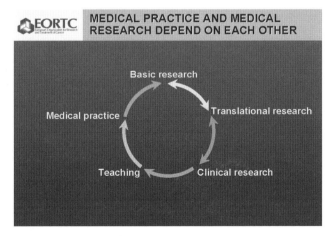

Whenever a new gene is discovered, the practical impact on patient care should only be expected after several years of painful, expensive and complex multidisciplinary clinical trials.

Clinical research is indeed a rather recent discipline and, at least in oncology consists of two different types of trials.

There are also trials to test and evaluate new agents leading to the registration of new medications. Those trials are usually conducted in partnership with the pharmaceutical industry or by the pharmaceutical industry alone.

Investigator driven trials and international cooperation

In oncology, there is a great need for non-commercial trials i.e. therapeutic strategy trials not aimed at the registration of drugs but of evaluating a multidisciplinary approach involving not only drugs (already on the market) but also surgical procedures and radiation oncology.

Investigator driven trials are essential and provide numerous opportunities for independent/objective evaluation but also to conduct large-scale trials that change clinical practice and to establish new state-of-the-art treatment with translational research components requiring the screening of large numbers of patients with similar tumor characteristics.

International cooperation is particularly needed for rare tumors. The initiation and the conducting of these investigator driven trials involve numerous partners including academic centers (the network of investigators) but also international and national networks, regulatory affairs agencies, ethics committees, cancer leagues, charities, other funding bodies, patient advocacy groups and, of course industry, whenever one of the therapeutic programs comprises a drug that is not yet registered.

Clinical research in oncology is also essential to identify ineffective and redundant treatments. In common and devastated malignancies, even a small improvement in survival will have a major impact on public health and health care budgets. In addition, in rare tumors, only multinational efforts will provide the opportunity to reach the required sample size. Therefore, there is a strong need to discourage small national-size trials that are inconclusive, unethical and that unnecessarily spend research funding without providing convincing data.

Academic clinical research focuses on clinically important and relevant questions and also has the advantage of providing independent control over tissue banking and translational research.

The European Clinical Trials Directive and its practical impact

All these efforts are under major threat due to the implementation of the EU Directive/2001/20/EC of the European Parliament and of the Council of 4 April 2001 on the approximation of the laws, regulations and administrative provisions of the Member States relating to the implementation of good clinical practice in the conduct of clinical trials on medical products for human use.

This Directive poses a number of important problems for academic research taking into account a lack of harmonization of national laws without improving either the quality of science or protection of the patients.

The costs related to the conducting of Pan European cancer clinical trials have almost doubled since 2004, which means that there is less critical research being carried out.

Figure 4
EORTC HQ staff and EORTC patients (1997-2007)

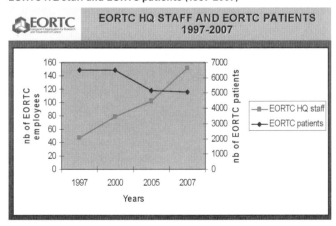

Figure 5
Workload: SUSARs

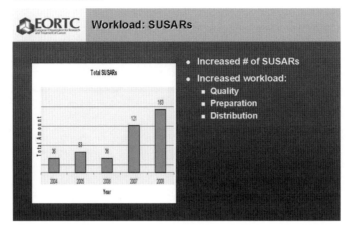

Tremendous efforts have been made by the academic community to cope with the increased regulatory burden, and indeed there should not be a two-tier framework for research (one for the pharmaceutical industry and another for academia), but there is a need to recognize the importance of the contribution of academic research to public health without jeopardizing the protection of patients in terms of rights and safety. Indeed, testing a compound that is registered and commercialized for more than twenty years has another dimension than testing a compound in a 'first in man' framework.

The voluntary contribution over the last decades of many clinical investigators has led to considerable improvements in health care in all disciplines of medicine, to the benefit of the patients but also to the visibility of scientific achievements of Europeans.

Currently, several steps have been taken within the European Commission to tackle the issue of non-commercial clinical trials. Meanwhile international networks such as EORTC have had to adapt their budget, procedures and staff numbers in order to be able to pursue international clinical trials without commercial aims and, at the same time, implementing translational research within complex and sophisticated 'clinico genomic trials' (see previous figures).

Achievements, dilemmas and challenges

There are numerous examples of the achievements of international academic clinical trials, for example, in the area of larynx preservation or less mutilating surgery for breast cancer, as a result of major pan-European clinical trials that would be extremely difficult to conduct today because of the increased bureaucracy and administrative workload.

Suboptimal funding of independent clinical trials is a disservice to patients and the scientific community; it promotes a brain drain even though Europe has the expertise of international clinical research and an excellent track record in the field.

These challenges offer opportunities to redefine the scientific community/patient organization dialogue in order to reinforce patients' interests and accrual in high quality cancer clinical trials. Cost containment is a major threat for pan-European translational and clinical research. Today, less than 5% of cancer patients benefit from clinical trials because of the lack of access to trials' methodology and infrastructure in most hospitals.

In the UK, faced with similar difficulties, one should stress the beneficial effect of a dramatic increase of financial support and human resources for cancer research, which has led to a major increase of participation of patients in clinical trials during the last five years with the support of the NHS and Cancer Research UK (a charity).

This means that there is a major need in Europe for appropriate clinical research infrastructures in hospitals and also for coordinating centers facing increased difficulties in relation to workload and pressures from health care administrations and hospital budgets.

There is a need to improve the training of health care professionals in the methodology of translational and clinical research and also to promote the independence of investigators by increasing structural and long-term funding to support the investigators and clinical research infrastructures. Although there is a lack of strong international networks, the existing ones provide tremendous support to investigators still willing to conduct independent international clinical research.

Particularly in oncology, there is a rapidly increasing portfolio of new compounds with new molecular targets as a result of discoveries made in basic research.

There is also a need for innovation in the design of the studies and in analysis of clinical trials including translational research components.

Therefore, cooperation between translational laboratories and networks of clinical investigators is of paramount importance if we want to promote our capacity for medical excellence in Europe and if we are convinced that clinical research is not a luxury.

Finally, there is a need for a new model of collaboration and true partnership between industry, policy makers, academia, charities, patient organizations and public funding. Such an initiative will require permanent dialogue between all stakeholders for the benefit of patients and the competitiveness of Europe in science and health care budgets.

Figure 6
Cancer clinical trial in the 21st century

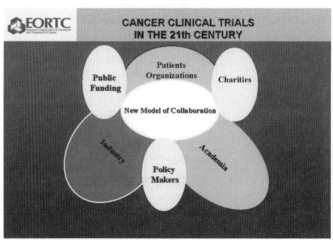

Future perspectives

If Europe is to remain competitive and provide new cancer treatments to patients, then there is a need to maintain and develop its clinical and translational medical research excellence. Investigator driven trials must be supported, the current legal framework should be revised and a new model of collaboration between research stakeholders is clearly needed. Only large international clinical research may change practice rapidly by avoiding unnecessary duplication at national level. EORTC is committed to providing its expertise to all investigators interested in collaboration to further improve the prognosis and the management of European citizens with cancer. This is particularly crucial at this moment since clinical trials are much more complex than ever, requiring screening of large numbers of patients in order to recruit sufficient individuals with the relevant tumor markers, taking into account 'clinico' genomic aspects and targeted individualized therapies.

More information about the EORTC

EORTC role and mission
The EORTC is a unique pan-European non-profit clinical cancer research organization established in Belgium in 1962, which operates as an association under Belgium law.

The original founding members of the EORTC were European pioneers of high quality multidisciplinary cancer research led by the Belgian Professor Henri Tagnon. The EORTC mission today remains the same. The EORTC develops, conducts, coordinates and stimulates high quality translational and clinical trial research to improve the standards of cancer patient care. This is achieved through the development of new drugs and other innovative approaches, and the testing of more effective therapeutic strategies, using currently approved drugs, surgery and/or radiotherapy in clinical trials conducted under the auspices of a vast network of cancer researchers.

The EORTC research network of more than 2,500 scientists and oncologists from 32 countries and over 300 university hospital facilities, ensures the timely passage of experimental discoveries into state-of-the-art treatments by reducing the time for research breakthroughs to reach clinical use.

EORTC research spans the entire spectrum from translational and drug development research to large, prospective, multi-centre, phase III clinical trials that evaluate new cancer therapies and/or treatment strategies as well as patient quality of life. Approximately 50 EORTC protocols are permanently open to patient recruitment and each year more than 5,500 new patients – over 85% from within the EU – receive their cancer treatment as part of an EORTC clinical trial protocol. A further 30,000 patients continue to be followed on a yearly basis and the EORTC clinical study database now contains outcome data for over 160,000 cancer patients.

Figure 7
EORTC clinical trials, 2000-2007

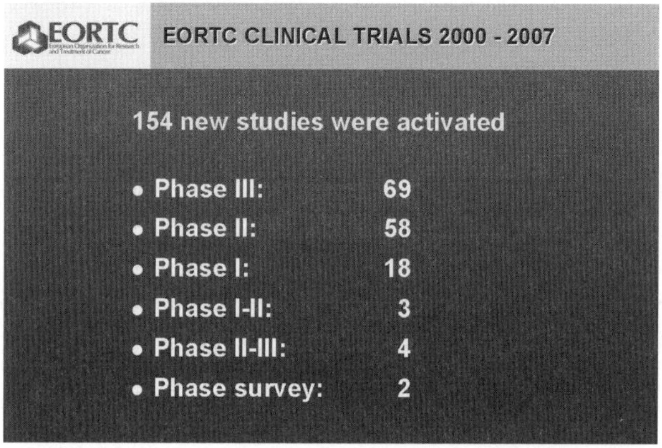

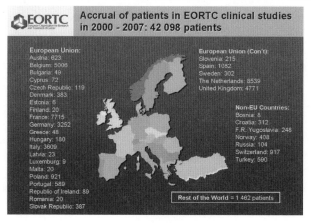

Figure 8
Accrual of patients in EORTC clinical studies in 2000-2007: 42,098 patients

Alongside the EORTC scientific and clinical program, the EORTC collaborates with a number of international organizations, including the US National Cancer Institute, the US Food and Drug Administration, the US Office for Human Research Protection of the National Institutes of Health, the European Medicines Evaluation Agency and many other national and international research groups worldwide.

The EORTC Headquarters, located in Brussels, Belgium, was established in 1974 to coordinate all EORTC scientific activities and related legal and administrative issues. The Headquarters Director General implements all EORTC strategies and policies as defined by the Board and leads a staff of 160 professionals representing 15 nationalities. The EORTC Headquarters' team works within a matrix organization of cross-functional departments, offices and units to provide full logistic and scientific support for all EORTC clinical trials and translational research projects.

The Headquarters' staff consists of medical doctors, statisticians, data managers, project managers, quality of life specialists, health care professionals, computer specialists, research fellows and administrative personnel. The Scientific Director oversees all medical and scientific activities related to prospective, newly proposed and ongoing EORTC clinical trials. The Director of Methodology-Operations manages all departments and units involved in the operational aspects of clinical trial data acquisition, analysis and quality assurance. The EORTC staff carries out specific research projects in the fields of statistics, quality of life, quality assurance and information technology. Between the years 2000 and 2007, EORTC investigators and the Headquarters' team published a total of 1,084 peer-reviewed medical and scientific papers.

EORTC Research Achievements

Major EORTC research successes and achievements in the past 45 years have contributed to:

- Improved survival for testicular cancer, childhood acute leukemia and lymphoma, adult Hodgkin's and non-Hodgkin's lymphoma, and gastrointestinal tumors (GIST).

- Improved standards of care for glioblastoma, melanoma, breast and colorectal cancer.
- The practice of organ preservation, minimal and non-mutilating cancer surgeries.
- The use of high-precision three-dimensional conformal radiation therapy techniques.
- The development of innovative complex therapeutic strategies and targeted agents.
- Improved treatments for severe and life-threatening cancer therapy-related infections.
- Improved patient quality of life, supportive care and better tolerated therapies.
- The development of outpatient cancer treatment strategies.
- The development of guidelines, quality of life questionnaires and prognostic calculators that aid the diagnostic and treatment decision-making process.

ACKNOWLEDGMENTS

Françoise Meunier wishes to thank Stéphane Lejeune and Diane van Vyve for their contribution to this article.

SELECTED READING

Annals of Oncology, J Ferlay & Al, *"Estimates of the cancer incidence and mortality in Europe in 2006"*, 18:581-592, 2007

Directive 2001/20/EC of the European Parliament and of the council of April 2001 L121/34 – Official Journal of the European Communities, 1.5.2001

Additional information on the EORTC website: www.eortc.be

BIO

Françoise Meunier, Director General of the European Organisation for Research and Treatment of Cancer (EORTC), received her medical degree (1974, summa cum laude) from the Université Libre de Bruxelles (ULB) and completed her research fellowship at the Memorial Sloan-Kettering Cancer Center in New York in 1977-1978 (Fulbright award). She holds a Master Degree from the ULB in both Medical Oncology (1976) and Internal Medicine (1979), and earned her PhD (Agrégé de l'Enseignement Supérieur) in 1985 at the ULB. She is also certified as a Pharmaceutical Medicine specialist by the Faculty of Pharmaceutical Medicine in the UK as well as in Belgium and has been a Fellow of the Royal College of Physicians of the UK since 1994. Françoise Meunier has led the coordination and administration of all EORTC activities since 1991 with the mandate to promote the EORTC as a major European organization in the field of oncology. As Director General, she is responsible for the organization of scientific activities, public relations and medium-term EORTC strategy as defined by the EORTC Board. Before joining the EORTC in 1991, Françoise Meunier was Head of the Infectious Disease Department at the Institut Jules Bordet in Brussels, Belgium and her personal area of research included mainly invasive fungal infections in cancer patients. She has published over 150 peer-reviewed articles and is a member of numerous international oncology scientific societies. She is a member of the Belgian Royal Academy of Medicine (Académie Royale de Médecine de Belgique) as of 2006. Françoise Meunier was awarded the Belgian Laureate "Prix Femmes d'Europe 2004-2005".

OPEN SOURCE DRUG DISCOVERY: THE TROPICAL DISEASE INITIATIVE[1]

SARA ENGELEN & MARC A. MARTI-RENOM

Inspired by the convergence of biology and computer science and the worldwide popularity of the grassroots approach of open source software projects during the nineties, Stephen M. Maurer, Arti Rai and Andrej Sali came up in 2004 with the idea to launch a web-based, bottom-up, decentralized initiative to discover new drugs. In 2006, this got them into the CNN Money.com's 50 Who Matter-list of the top 50 cutting-edge innovators who are setting today's business agenda. Considering the low number of new therapeutics in recent years and global concern about declining productivity and innovation in pharmaceuticals, this initiative is a call to join forces to tackle medical challenges that the blockbuster model cannot address economically, such as tropical diseases.

More eyeballs

The landscape of innovation doesn't exist only of patents and copyrights. The ultimate proof of concept for this was delivered by Linus Torvalds some 15 years ago. As opposed to conventional trajectories in software development, he dramatically re-sketched the R&D dynamics flow by making the source code of a software operating system available to the general public on the Internet. By creating an online community of software programmers based on the *"given enough eyeballs, all bugs are shallow"*[2] principle, he intro-

---

1 This article summarizes a talk given by Marc A. Marti-Renom, Structural Genomics Unit, Prince Felipe Research Center, Valencia at the CROSSTALKS Science & Industry Lunch on "New Medicine Research Collaborations", held in April 2008 at VUB Brussels and presents some additional reflections from existing literature on open source drug discovery.
2 Eric Steven Raymond is a computer programmer, open source software advocate and author of *"The Cathedral and the Bazaar"* (1997), in which he launched the famous expression that *"given enough eyeballs, all bugs are shallow"* – the more widely available the source code of a computer program is for public testing, the more rapidly bugs will be discovered.

duced a faster, cheaper, and socially *engaged* way of delivering products to a new generation of consumers and developers. Today the movement's crown jewel, Linux, has become a credible alternative to mainstream operating systems.

To the surprise of many non-believers, protective intellectual property strategies and rigid publication regimes no longer prevent a growing cohort of scientists from disclosing their findings to the public domain. During the past 10 years, a new breed of organizations, so-called public-private partnerships, has been adapting the open source R&D model to enable scientists to freely solve problems of common interest across organizations, disciplines and borders. The international effort to sequence the human genome, for instance, irrespective of the massive top-down government involvement, resembled an open source initiative and inspired other initiatives such as the SNP Consortium, the Alliance for Cellular Signaling, BioForge, GMOD and Massachusetts Institute of Technology's BioBricks.

Why would anyone bother contributing his or her knowledge without any financial compensation? As open source's success story shows: because some people simply value expertise, technical and intellectual challenges and community feeling more than money. Expertise is volunteered to satisfy idealism or curiosity, seek new challenges, improve skills, build a reputation or enhance careers. Another advantage of open source tactics is that sharing information can foreclose new uses or enhancements of existing products or services by outside researchers.

It's common practice that physicians share their ideas and experiences informally to uncover novel uses for existing medicines. Many medications are approved for one purpose, but are regularly prescribed for another, 'off-label' use. However, off-label drugs rarely go through the formal process for other uses because the cost of regulatory approval is so high. Avastin for instance is an internationally debated case. Currently there are three drugs that have been used in the treatment of a particular eye disease, wet macular degeneration: Macugen, Lucentis and Avastin. Although Avastin is much cheaper than Macugen and Lucentis and already available on the market as an approved treatment for colon and rectal cancer, it has never been approved by the EMEA as eye medication. This is a lost opportunity for society because the effectiveness of the treatment is not formally evaluated, social security invests money that could be better spent elsewhere and patients don't get the treatment that would really benefit them, be that therapeutically or financially. This is a persisting situation, much to the astonishment of many ophthalmologists, who have to inject the drug at their own risk and hence face huge legal claims if something goes wrong.

Eric von Hippel, Head of the Innovation and Entrepreneurship Group at the MIT Sloan School of Management, has been investigating how secondary uses for drugs are discovered, with a view to harnessing doctors and patients to record data. His proposal is to decentralize the process of obtaining data on the off-label use, by collaborating with volunteer doctors and patients. By defraying costs in this way, it might then be possible to obtain regulatory approval. Because the drug has already been approved, it has passed first phase tests for safety, which don't have to be repeated. Second and third phase drug approvals test for efficacy and side effects – the very areas where getting formal approval for off-label use is sensible[3].

The idea is, in effect, an open source clinical trial, which amongst so many other similar proposals in the health care pipeline, tries to introduce a new dynamic into a worn-out business model, adding substantial arguments to the discussion on incremental innovation in medicine and how to reward it. This article focuses on one such remarkable initiative on open source drug discovery, The Tropical Disease Initiative (TDI), which might eventually be matched by open source clinical trials down the road.

A new way of doing things differently

"We are so used to patents that we forgot ways to discover drugs in the public domain, and we need to rediscover them" – Stephen Maurer

With regard to drug development, in the same way that programmers find bugs and write patches, biologists look for proteins ('targets') and select chemicals ('drug candidates') that bind to them and affect their behavior in desirable ways. Medicines – like software – consist largely of information: in both cases, research consists of finding and fixing tiny problems hidden in

---

3 *An open source shot in the arm?* Technology Quarterly. Published online at Economy.com at www.economist.com/science/tq/displayStory.cfm?story_id=2724420

an ocean of code[4]. Open source's chief benefit is to cross-fertilize minds and tap creativity quickly and cheaply, and on a scale that is beyond the reach of scientists working in the ivory towers of academia and the corporate moats of industry.

What would open source drug discovery look like? Knowledge-based work requires lots of intelligence and intuition, but little infrastructure. Examples include identifying targets, understanding metabolic networks, and designing clinical trials or computerized disease models – scientific work that is excellently suited to the computer. It is about scientists leveraging each other's ideas, and using tools to gain deeper insights that might lead to breakthroughs.

After documented research on existing initiatives to pool knowledge in the field of biotech and pharmaceuticals, Stephen Maurer (University of California School of Public Policy, Berkeley), Arti Rai (Duke University School of Law, Durham) and Andrej Sali (Dept of Biopharmaceutical Sciences and Pharmaceutical Chemistry, San Francisco) launched a proposal in 2004 that described an open source drug discovery project. With the aid of Ginger Taylor, founder and executive director of the Synaptic Leap – an organization dedicated to providing a network of online communities that connect and empower scientific and medical researchers to conduct open source style research – they developed online tools that enable scientists to collaborate without having to actually meet. Similar to an open source software project, they aimed for a bottom-up self-organization among researchers themselves: individual pages would host tasks like searching for new targets, finding chemicals to attack known targets, and posting data from related chemistry and biology experiments. Volunteers could use chat rooms and bulletin boards to announce discoveries and debate future research directions. Over time, the most dedicated and proficient volunteers would become leaders. The final development of drug candidates would be awarded to a laboratory based on competitive bids. The drug itself would go into the public domain, for generic manufacturers to produce[5].

---

4   Maurer S. (2004). *Finding Cures for Tropical Diseases: Is Open Source an Answer?* PloS Medicine, Volume 1(3), p.33-37.
5   idem

In 2005 the website 'The Tropical Disease Initiative' was launched and as Marc A. Marti Renom explained in April 2008 during a CROSSTALKS Science & Industry Lunch on 'New Medicine Research Collaborations': *"By 2006 we were gaining some momentum, and Maurer and Sali got into this 50 people who matter list, that CNN and Money.com put together every year about people who actually think ahead and care about new ways of doing things. Between 2004 and 2007 a lot of students, scientists and institutions would come to us to express their interest, but the problem was that we didn't have a common ground – I like to call it a kernel, similar to what was so important for the Linux community in the nineties and what brought all those very diverse communities of developers together. We believe that somehow we might have this kernel ready after three or four years working within the framework of the Tropical Disease Initiative."*

[233]

Finding a niche

Looking for a location along the drug discovery pipeline where patent driven R&D is compatible with open source methods, open source research could open up two areas in particular. The first is that of non-patentable compounds and drugs whose patents have expired. They receive very little attention from researchers, because there would be no way to protect (and so profit from) any discovery that was made about their effectiveness. As already mentioned, lots of potentially useful drugs could be sitting under researchers' noses, but there is a lack of incentive to start looking.

The second area where open source might be able to help would be in developing treatments for diseases that afflict relatively small numbers of people, such as Parkinson's disease, or that are found mainly in poor countries, in tropical and underdeveloped areas where needs are huge but funds are scarce. And that's where the Tropical Disease Initiative comes in. Very few newly developed drugs are aimed at tropical, 'neglected' diseases. In contrast with northern 'medically well-catered' countries, where heart disease and cancer are the main causes of death, a DALY (Disability Adjusted Life Years) chart points out that in developing countries, tropical diseases such as dengue fever, schistosomiasis and malaria still impose a heavy burden on society. *"Malaria in particular is bad"*, stated Marc A. Marti-Renom, *"the global burden of malaria expressed in DALYs is 46 million, which would mean that if malaria isn't completely cured, the entire population of a country the size of Spain wouldn't ever be productive"*.

Why do so many die? As TDI-founder Stephen Maurer and his colleague initiators Arti Rai and Andrej Sali explain in their 2004 proposal, the reasons are more economic than scientific, since grants and patent incentives were never designed with tropical diseases in mind. Traditional pharmaceutical companies cover their R&D costs by selling patented products. This strategy fails in the developing world, where would-be consumers are often penniless, and tropical diseases are as such 'unprofitable'. Reform proposals asking governments and charities to subsidize developing country purchases at a guaranteed price might hold the key to a solution, but don't subvert the patent system. Another approach is based on charities that create non-profit venture-capital firms (Virtual Pharmas). Virtual Pharmas collect a portfolio of promising drug candidates and then outsource them through contracts with commercial and academic partners.

Figure 1

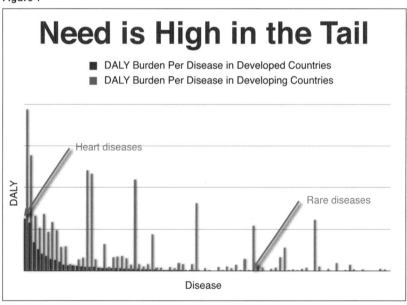

Source: Disease data taken from WHO, World Health Report 2004

DALY is not a perfect measure of market size, but is certainly a good measure for importance. DALYs for a disease are the sum of the years of life lost due to premature mortality (YLL) in the population and the years lost due to disability (YLD) for incident cases of the health condition. The DALY is a health gap measure that extends the concept of potential years of life lost due to premature death (PYLL) to include equivalent years of 'healthy' life lost in states of less than full health, broadly termed disability. One DALY represents the loss of one year of equivalent full health.

But, anticipates Maurer, Virtual Pharmas are insufficient, as they face three important obstacles. First, assessing expected R&D costs to get funding for a project is very difficult – one needs to understand what a product costs in order to negotiate the best price – with estimates ranging from $100 to $500 million per drug. Second, development pipelines will run dry without more upstream research, especially on genomic insights. And third: in general tropical disease research is badly funded and rigid cost containment is urgently needed.

The Tropical Diseases Initiative aims to keep the pipeline of virtual pharma full. Five distinctive target areas are defined, in which scientific discoveries will add up to a continuous work in progress – referred to as the kernel: in silico (or computer-based) drug discovery and chemistry, stem cell lines, and more ambitiously, phase IV trials, and phase I to III trials.

Stimulating serendipity

The life cycle of a new drug takes about eight to 12 years from discovery to registration, in which one's freedom to operate declines with the passage of time, since IP-rules start to apply. In silico drug discovery and chemistry, which were the focus of Marti-Renom's CROSSTALKS presentation and which today are TDI's core activity, have to be situated in the very beginning of the drug discovery pipeline, in the pre-lead area, that still has very open characteristics. As drug discovery is a highly serendipitous process, and as *"chance favors the prepared mind"* (Pasteur), the Tropical Disease Initiative wants to streamline and optimize the information supply and prepare several minds alike with an eye to finding new targets.

The increasing accessibility and availability of biological databases, cheaper and faster software and computers, and the Internet enable more people to work simultaneously on more projects at a lower cost. Today, models based on computational biology and chemistry can predict the toxicity and drug likeness of a molecule and reduce the number of random walks in the discovery process. By increasing the success rate of the initial set of leads by isolating potential breakthrough molecules and hence shortening the times of target and lead identification, Marti-Renom continued, the initiative's approach can generate considerable impact.

As Marti-Renom pointed out, "It is striking that of the 361 new molecular entities approved by the FDA between 1989 and 2000, 76% targeted a preceding well-drugged domain and only 6% targeted a previously un-drugged domain. The question of course is where and how to find those new drugs. Based on computational biology, we try to identify new possibly interesting targets, and that's what we call the kernel; the information we put out there in the public domain, so people can start playing and exploring. This is a computational approach, where the stress is on the predictive character, and the more people look at it, the bigger the chance some real hits are found. A growing interdisciplinary network of scientists must move the initiative further over time. (...) Our work consisted in taking 10 genomes from pathogens that cause tropical diseases, such as leprosy, tuberculosis or malaria. Those 10 genomes contain about 60,000 genes, from which we predicted reliable protein structure models for about 16,000 genes. Based on further estimations, from those 16,000 genes, 3,500 genes are likely to bind a small molecule, from which 297 are predicted to bind drug-like molecules and 143 are predicted to bind known drugs. This analysis, thus, results in an initial kernel for drug discovery that people can take over to do virtual screening and leads to optimization. If there is a 'hit' with a particular compound, it can be extracted from the database for further in silico testing and then taken into the laboratory for experimental validation. This would help in achieving the goal of getting new medicines to those who need them, at the lowest possible price."

Cost containment

As such, TDI can be a great boost to the efforts of Virtual Pharmas in helping to contain the costs of discovering drugs. First, similar to the open source software community, TDI relies on a horizontal innovation network and asks volunteers to donate their time (and any patentable discoveries) to the collaboration. Instead of financial incentives, volunteers are offered non-monetary rewards, such as ideological satisfaction, the acquisition of new skills, enhancement of professional reputation, and the ability to advertise one's skills to potential employers.

Will universities and corporations really let their people volunteer? "They have little to lose", is TDI's argument, "as the value of their intellectual property depends almost entirely on US and European diseases". Ultimately, the initiative expects these institutions to also donate data, research tools and other resources to make TDI stronger.

Secondly, since TDI fosters open source licensing and avoids working with patented data – *"which short circuit competition by giving the owners the legal right to prevent others from using"* – an entire new field of research competition is created and drug candidates are available to anyone who wants to develop them. Sponsors are expected to exploit this advantage by signing development contracts with whichever company offers the lowest bid: Virtual Pharmas can choose the best candidates to develop the drugs. And it is argued, this kind of contract makes good business sense, as risks and benefits of R&D are shared, and the absence of patents should keep prices low once drugs reach the market.

Figure 2
TDI Flowchart

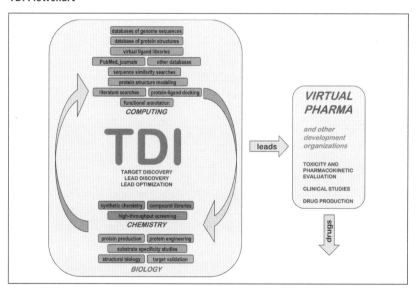

Conclusion

Can open-source drug discovery revitalize medical research and genuinely foster innovation? It's a fact that many new business alliances in pharma and biotech companies mirror the synergetic features of the open source model in their attempts to implement the open innovation model. Since open source networks are richer in 'weak links' (loose relationships), they are said to bring novel ideas whereas strong links tend to reinforce orthodoxies. The question remains, however, whether corporate alliances will allow for so much 'undefinedness' to be economically feasible. The Tropical Disease Initiative is a proof of principle that there is indeed a place for collaborative and proprietary research in drug R&D, just as in software. However, like all open source drug discovery collaborations, TDI is an experiment, and whether open source drug discovery becomes a transformational force or remains a non-threatening niche depends on how it ultimately performs against traditional pharmaceutical R&D.

Reviewing some of today's attempts to organize open source biology collaborations, Maurer's general conclusion is: *"The stakes are high. We have seen that open source is often a plausible strategy for reducing drug development costs and making new medicines affordable; that it offers increased transparency for funding agencies trying to decide which early stage drug candidates to invest in; that it may allow regulators to reduce the reporting requirements that help make late-stage drug discovery expensive; and that its ability to mobilize volunteers offers a key advantage to cash-strapped neglected disease programs. Perhaps more importantly, open source is the first fundamentally new innovation mechanism since patents and copyright appeared four centuries ago."* [6]

If open source drug R&D takes hold, what will probably emerge is not the replacement of one model by another, but an ecology in which big pharma, biotech and collaborative research compete and collaborate at the same time, feeding off each other synergistically, while moving towards therapies along their own distinctive paths [7]. The Tropical Disease Initiative provides a way to leverage capabilities to tackle unmet medical needs, such as the diseases

---

6   Maurer S. (2008). *Open Source Drug Discovery. Finding A Niche (Or Maybe Several)*. UMKC Law Review 76:2.
7   Munos B. (2006). Can open-source R&D reinvigorate drug research? Nature Reviews Drug Discovery. Advance Online Publication, published online 18 August 2006.

of poverty, orphan diseases and niche markets. Nevertheless, just as putting ingredients into a vat does not necessarily cause them to react, connecting smart people online does not guarantee they will produce anything valuable. In both cases, a catalyst is needed. To succeed, open source R&D will need structural funding and the expertise of drug R&D, which today resides overwhelmingly in the pharmaceutical industry. There might be many volunteers, but they must be shepherded towards a goal. Such stewardship is a core competency of pharmaceutical companies[8]. And, as Marti-Renom concluded, *"good science eventually generates its own funding"*.

ACKNOWLEDGEMENTS

A word of gratitude goes to Rita De Doncker, ophthalmologist, who provided information on the Avastin case and to Erik Verschueren and Thomas Crispeels for their valuable comments. MAM-R acknowledges support from the Spanish Ministry of Education and Science (BIO2007/66670).

FURTHER READING

http://www.tropicaldisease.org
http://www.thesynapticleap.org
http://www.collaborativedrug.com
http://sgu.bioinfo.cipf.es
http://www.salilab.org

ILLUSTRATIONS

All illustrations were taken from Marc A. Marti-Renom's presentation at the CROSSTALKS workshop in April 2008.

BIOS

Marc A. Marti-Renom is a computational biologist and head of the Structural Genomics Unit of the Bioinformatics Department, Prince Felipe Research Center. He conducts research on the improvement of the accuracy of protein 3D models. He is a founding member of the Tropical Disease Initiative. Using an open-source research and development model, the Tropical Disease Initiative seeks cures for tropical 'orphan' illnesses such as malaria, dengue fever and African sleeping sickness. TDI aspires to combine the efforts of hundreds of volunteer researchers from around the globe, focusing on the application of computational biology and chemistry on drug discovery. Relying on a distributed, collaborative, transparent process of biomedical development, this bottom-up approach might take on health challenges that pharmaceutical corporations have determined as unprofitable.

Sara Engelen (°1978) graduated in Communication Science at K.U.Leuven and obtained a Master in Applied Audio-Visual Communication Sciences. She has worked on the editorial staff of regional television stations and as a production assistant at various international cultural events. A freelance copywriter, since 2004 she has been part of the Technology Transfer Interface team and co-developer of the CROSSTALKS network at Vrije Universiteit Brussel.

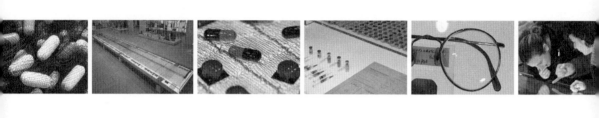

EPILOGUE

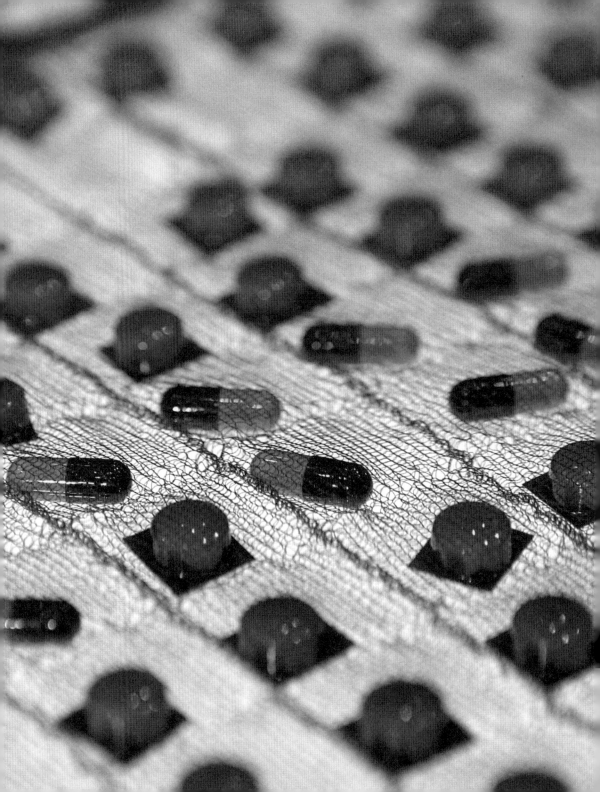

CRADLE TO GRAVE

A PROJECT AND ART INSTALLATION ON PERMANENT DISPLAY
AT THE BRITISH MUSEUM, LONDON[1]

SUSIE FREEMAN, DAVID CRITCHLEY & LIZ LEE

In 2003, Pharmacopoeia received a major commission from the British Museum, leading to the work 'Cradle to Grave'. The installation tells the story of an average man and woman told through the medication they have taken in their life and accompanied by photographs, documents and objects. 'Cradle to Grave' sheds a particular light on a person's medical history from a medicines' point of view. Depending on how old the onlooker is, the installation likewise projects a glimpse into a possible future and a thorough reflection on that.

Pharmacopoeia is a medical-art collaboration between the artists Susie Freeman and David Critchley and the family doctor Liz Lee. Over the last ten years we have created a body of work that explores different aspects of health and ill health. Most of our work reflects attitudes and health beliefs that are common in the UK, but we have also worked on projects in other European countries and in Africa. Central to our work are active pharmaceuticals that we buy from pharmacies using private prescriptions issued by Dr Lee. In this way we access real drugs that are not available to ordinary members of the public unless they are prescribed. In some artworks we also use 'over the counter' medicines that can be purchased without a prescription. The only drugs we do not use for legal reasons are 'controlled' drugs such as morphine.

---

1   The contemporary art installation 'Cradle to Grave' dates from 2003 and is on display in the Wellcome Gallery at the British Museum, London. See: www.cradletograve.org

The pills and capsules and often their packaging, are incorporated into fabric by a process known as 'pocket knitting'. By using a fine nylon yarn small solid objects such as pills are captured in pockets in order to create large flexible fabrics.

In 2001, the British Museum commissioned Pharmacopoeia to contribute to their new gallery of ethnography 'Living and Dying'. We were asked to make a piece of work that reflected how people in our own Western society respond to sickness and ill health and how we also strive to promote and preserve our sense of wellbeing. The art installation we created is called 'Cradle to Grave'. It focuses on the Western biomedical approach to ill health with its reliance on medicines, which we take in ever increasing amounts as we move from birth, childhood and adulthood into old age and eventually death. Within the gallery this is contrasted with a number of other societies from the Western Pacific, Nicobar Islands, Native North America and Bolivian Andes who all invoke the help of spirits or Gods to protect them from harm and to cure them of sickness.

Cradle to Grave is a 14-meter long installation that runs down the middle of The Wellcome Trust Gallery at the British Museum. Its central theme is the dominance of the biomedical approach to health and illness within Western societies, here focused mainly on the use of medication. Two central 'Pill Diaries' made out of pocket knitted fabric, document and reveal all the medicines prescribed to one woman and one man during their lifetime in the UK today. The pills are laid out in the exact sequence in which they would be taken. As most people also employ a variety of other strategies to promote their sense of wellbeing and to combat ill health, several linked narratives explore these complementary themes. These are provided by a series of objects, documents, and personal photographs that run along either side of the Pill Diaries. Together they reflect the ways in which people deal with sickness and try to secure wellbeing in the UK at the beginning of the 21st century.

Research and methodology

Having received the commission, we started looking at national and international mortality and morbidity data to ascertain major causes of illness and ill health. Strokes and heart disease are the most common causes of deaths both in the UK and worldwide. Other important conditions contribute to morbidity (ill health) rather than mortality (death). For example, depression is rated as one of the top five causes of morbidity worldwide. From this data we began to map out the kinds of illness we wanted to include in the work.

We then looked at the national prescribing figures. The numbers are shocking. For example, currently there are over 40,000,000 prescriptions for antidepressants issued in the UK every year. We discovered other interesting facts, such as on average everyone in the UK takes one course of antibiotics every two years and that we spend more treating indigestion than we do treating cancer. This information provided us with a framework within which to construct the narratives of 'Everyman' and 'Everywoman'.

Both contain over 14,000 drugs which is the current estimated average prescribed to every man and woman in the UK in their lifetime. It should be emphasized that these are only prescribed medicines and so over the counter medicines or pills such as vitamins or minerals are not included. For example, most people do not get a prescription for painkillers when they have a headache; they buy paracetamol over the counter at the pharmacy. Many people take multivitamin tablets or antioxidants daily, which are not prescribed by a doctor. Many take indigestion tablets or laxatives, all of which they buy without a prescription. Although they may make a significant contribution to our health, none of these are included in Cradle to Grave.

Pharmacopoeia's work is always based on real people's records and on actual medication. This presented us with a problem. We could not simply use the entire medical record of an 80 year old. After all, they were born before the development of most of our current drugs and although it might provide an accurate historical record, it would have little relevance to today's population. In the end we took the pragmatic decision to create a composite 'Everyman' from the real medical prescribing record of four different males and an 'Everywoman' from four different females.

We were able to use real and accurate prescribing data because as a practicing family doctor in the UK, Liz Lee has access to the computerized prescribing records of the 13,000 patients registered at her practice. In the UK there is a reliable system of transfer of medical records when a person moves from one medical practice to another. Consequently nearly everyone's medical record accurately documents their medical history and prescribing from birth onwards.

Selection of pills

From the 13,000 records, we selected a 20 year old man who had had a number of common medical illnesses. We documented everything that he had been prescribed since he was born.

We also noted his childhood immunizations. We then chose the record of a 40 year old man and documented everything he had been prescribed from the age of 20 to 40. Again these were mostly treatments for common conditions. We ensured that the two patients' medical history 'fitted together' well by selecting suitable patients. For example, they both suffered from hay fever and asthma and this allowed us to ensure some continuity between the two sections of the narrative. We then repeated this process for a 60 year old and completed the record with the medication history of an ex-smoker with a bad chest and high blood pressure who died of a stroke at the age of 76, currently the average life expectancy of a man in the UK. Themes such as chest disease, indigestion and back pain were carried through all the four ages in order to represent the imaginary subjects of our piece.

The same amalgamation of four reasonably well matched patient prescribing records was used for the woman's diary. She took contraceptives regularly when she was young. In her twenties and thirties she had two children and one miscarriage, and was treated for an episode of postnatal depression. Later she was prescribed hormone replacement therapy for menopausal symptoms. After a mammogram in her fifties she was diagnosed and successfully treated for breast cancer. In her seventies she took increasing numbers of painkillers to treat arthritis, which eventually led her to have a hip replacement. Like an increasing number of people she was also diagnosed with diabetes. At the end of the diary she is still alive and reasonably healthy aged 82. In 2003 when we made Cradle to Grave the average life expectancy of UK women was just over 82 years.

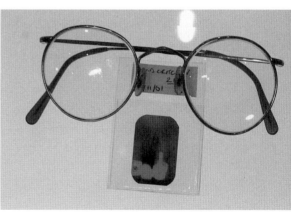

Selection of objects and photographs

The pill narratives provide the central structure for the piece but reveal only part of the complex strategies we employ in order to maintain a sense of health and wellbeing. Some of this complexity is captured in two other narrative strands. Running on either side of the pill diaries are personal objects, documents and medical artifacts that relate to daily life. Interspersing these are groups of photographs with captions written by their owners, tracing typical moments in real people's lives. The photographs are drawn from the albums of family, friends and colleagues. We invited a wide spectrum of people to submit photographs that they felt particularly illustrated their own personal experience of health and ill health. The response we got demonstrates very clearly that maintaining a sense of wellbeing is much more complex than just treating periods of illness. Among other things the photographs reveal that it is about family and community, work, weddings and funerals. It is about eating and drinking and smoking and dancing. It is about our relationship with nature. It includes sadness and suffering and loss.

The objects are more diverse still. These were selected by the artists in order to reflect the complexity and sophistication of our thinking and actions. They included choices we can make about healthy living as opposed to risk taking behavior. An apple to illustrate a healthy diet, condoms for protection against sexually transmitted disease, a glass of red wine, which is protective against heart disease, but in excess can damage our social and physical functioning. Conflicting feelings about 'healthy behaviors' are addressed by the inclusion of an ashtray filled with fag butts suggesting the dangers of smoking while the photographs acknowledge the pleasure and sociability associated with smoking.

Medical artifacts fill the gap created by our tight focus on medication. The contribution of technology and surgery to the biomedical approach to health is represented by x-rays, a pregnancy scan, a mammogram showing a breast cancer, and a prosthetic hip joint. The existence of a National Health Service, which undertakes to provide care that is free at the point of delivery to all residents in the country, is important to citizens' sense of wellbeing. It is represented by a blood donor collection bag and a long service enameled badge. In the UK, ordinary people donate blood as volunteers. They are not paid for their altruism but instead are rewarded for multiple donations with a blood donor's badge. Acupuncture needles and homeopathic medicines represent complementary therapies and a bible is included to acknowledge the importance of faith to many people. Finally there is the documentation, which in the case of the birth certificate, acknowledges our arrival into society and the death certificate, which marks our departure.

The pills

Cradle to Grave focuses on ordinary people suffering from the common ills of our society. Most of the medicines present in the two pill diaries are prescribed either for the primary or secondary prevention of disease. Primary prevention is treatment taken before a disease has developed. For example, for some years the man takes antihypertensive medicine to treat his high blood pressure. High blood pressure is not itself a disease, but if you suffer from untreated raised blood pressure it increases your likelihood of having

a stroke or a heart attack. In spite of his years of treatment, he does have a heart attack at the age of 76. After this his pill regime changes to one of secondary prevention. This means treatment is now aimed at preventing a second heart attack. This includes continuing to control blood pressure, taking a drug to reduce circulating cholesterol and taking an aspirin to thin the blood. These actions together will statistically reduce his likelihood of a recurrence.

The woman represented has breast cancer treated with surgery. After this she takes a pill every day for the next five years to reduce the likelihood of a recurrence of the cancer. Again this is secondary prevention. When trying to get pregnant and in the first three months of pregnancy she takes the vitamin folic acid to reduce the chance of her baby being born with the condition spina bifida. This is primary prevention.

Some of the medication is prescribed in order to cure, for example, both take antibiotics to cure infections of the chest or the throat. As well as primary or secondary prevention and cure, many pills are taken to control distressing symptoms such as indigestion or the pain of arthritis. Most of the early sections of the woman's diary are dedicated to the control of fertility. Thousands of contraceptive pills are taken in order to prevent the 'natural' act of conception.

There is also evidence of the medicalization of ordinary life: of the menopause, of unhappiness, of obesity and of smoking addiction. These more controversial areas of treatment are perhaps more susceptible to prescribing fashions. For example, hormone replacement therapy (HRT) was widely used in the UK five years ago. But now there is new evidence about its harmful as well as its beneficial effects and if we were remaking Cradle to Grave in 2008, HRT would not be included.

Other evidence based changes have also taken place. In 2002, by the age of four, children in the UK were immunized against nine infectious diseases, but this has now increased to 10. There are new guidelines on the most effective treatment of high blood pressure and changing trends in the medicines used to treat childhood pain and fever. Because of the speed of change in prescribing patterns, the pill narratives in Cradle to Grave have already become a historical record.

Looking into the future

Some people, including doctors, are therapeutic nihilists; others are committed pill takers. Each individual's response to Cradle to Grave reflects not only this natural preference but also their personal experience of illness. The tendency is for younger people to say *"this is not relevant to me as I hardly ever take any medication"*. In this they are correct. Young people take very few prescription medicines and if they look more closely they will see that this interpretation is clearly reflected in the work. Age tags sewn into the margin of the 14-meter strip of fabric reveal that by the age of 20, the amount of pills representing the intake of an average man, is only two meters long. One of the astonishing aspects of Cradle to Grave is that it not only allows a 20 year old to reflect on their present and past state of health, it also asks them to look into their future. It is at the other end that the real pill taking starts. 'Everyman' takes as many pills in the last 10 years of his life as he has in his previous 66 years.

Cradle to Grave incorporates evidence of the medicalization of ordinary life. We take pills to treat unhappiness, obesity, smoking addiction, to control natural events such as the menopause and these are important issues that our society needs to debate. Perhaps even more importantly, Cradle to Grave demonstrates our commitment to the medicalization of old age. As the body begins to fail, we turn to pharmaceutically active chemicals to preserve and extend life. We minimize the suffering of old age by medicating it. But this does raise questions on the earlier years when we are not considering long-term health, nor being concerned with health itself and only reacting to acute and crucial situations.

In the end we are asked to consider the deeply complex relationship we have with prescription drugs. They are both wonderful and dangerous. They allow us to live longer, they allow us to suffer less, but they may also offer false promises of happiness and health and immortality that they cannot possibly deliver. In this they are more like the spirits and gods of other cultures than we care to believe.

BIOS

Susie Freeman and Dr Liz Lee began to work together as Pharmacopoeia in 1998 and in that year won a Wellcome Trust Sciart Award. David Critchley joined them in 2000 to develop the project into a touring exhibition comprising 21 pieces ranging from ball gowns and framed works to photographs and installations. For 10 years they have continued to collaborate in the production of art works connected with medical issues.

Susie Freeman is a textile artist who has exhibited widely around the world. Her work is in the form of clothes, fabrics and framed pieces constructed with a unique pocket technique. While sometimes purely decorative, her pieces often illustrate themes of an artistic, personal or medical nature. She studied Textiles and Fashion at Manchester Polytechnic and has an MA in Textiles from the Royal College of Art. Her work is in many public collections including The Victoria and Albert Museum, Wellcome Collection, Gallery of English Costume Manchester, Bergen Museum of Decorative Arts Norway, The British Council and Kwangju Textile Museum, South Korea.

David Critchley exhibited in 'The Video Show' at the Serpentine Gallery in 1975, was one of the organizers of the influential series of installations and media performances at 2B Butler's Wharf in the late 1970s and was a central figure in establishing and running London Video Arts – now the LUX and Luxonline. His video artworks have been screened in the UK, Europe, North America and worldwide. 'Trialogue' was screened as part of the Century of Artists Film and Video at Tate Britain. He taught video art at the Slade, Chelsea and at many UK art colleges, and continues to be involved in art education. He is currently working on new site-specific multi-media art installations.

Dr Liz Lee works as a family doctor in Bristol where she is a partner in a large urban general medical practice. Her areas of special interest are cancer and end of life care. She has written widely on the subject and her book 'In Your Own Time' published by Oxford University Press is a guide for patients and carers facing terminal illness at home. From 1988 to 1995 she was a forensic medical examiner for the Metropolitan Police working with victims of rape and assault. Until recently she was a member of the National Advisory Committee on Breast Screening and the South and West Regional Tumor Panel for cancer of the head and neck.

LIST OF PICTURES

Cover photo
'My Smoking Family' by Claire Collison
An image from 'Cradle to Grave' art installation by Susie Freeman, Dr Liz Lee and David Critchley, The Wellcome Gallery at the British Museum

© **Marleen Wynants**

Page 6:	ZEIT – Yard of Kunsthaus Tacheles at Oranienburger Straße, Berlin, August 2008
Page 14:	Rodin Statue (from the Burghers of Calais) – Rodin Sculpture Garden at Stanford University, April 2008
Page 26:	How long is now – Mauer Tacheles, Oranienburger Strasse, Mitte, Berlin
Page 40:	Oh nein – Advertising for a dating organization, Charlottenburg, Berlin, August 2008
Page 56:	Flea Market – Piazza Peranni-Papiret, Palermo, Sicily, November 2005
Page 64:	Last Public Smoker – MItte, Berlin, August 2008
Page 76:	I'm here to save you money – Haight Ashbury, San Francisco, March 2008
Page 98:	Jack – DVD shop in Kreuzberg, Berlin, August 2008
Page 114:	Motor biking – Valle Gran Rey, La Gomera, January 2006
Page 124:	Carnival – Murcia, Spain, February 2006
Page 142:	Hiking Trail – La Gomera, January 2006
Page 154:	Künstlergemeinschaft NEUES PROBLEM in KulturHaus Mitte, Berlin, August 2008
Page 166:	Street art near Mauerpark – Berlin, October 2008
Page 182:	Schöne Dinge – Mitte, Berlin, August 2008
Page 196:	Keep it simple – Van Morrison Advertising, San Francisco, April 2008
Page 214:	Listening – Crowd listening to local group Fairy at Open Air Arena in Mauerpark, Berlin, October 2008
Page 228:	Local street scene – Sal Rei, Boa Vista, Cap Verde, January 2007

LIST OF PICTURES

© Tom Lee and David Critchley

Page 244: Tamoxifen & Flucloxacillin
Page 246: Cradle to Grave at the British Museum
Page 251: Early Years
Page 252: Heart disease treatment
                Sports day
                Gertrude
                Example of objects – Mature years portrait
Page 254: Petra, Felix & Maddy
Page 257: Three viewers

INDEX

A
accountability *23, 172, 190, 191*
AIFA *59, 60, 62*
Alzheimer *28, 29, 31, 59, 62, 101, 162*
angiogenesis *67, 68, 69, 70*
assessment *15, 25, 41, 42, 43, 54, 209*

B
Bill & Melinda Gates Foundation *22, 190*
biomarkers *70, 71, 73*
biomedicine *29*
breakthrough medicines *35, 36, 126*
Busquin Act *172*

C
cancer *17, 28, 31, 33, 60, 65, 66, 67, 68, 69, 70, 71, 72, 74, 75, 101, 106, 109, 150, 178, 215, 216, 217, 218, 219, 221, 223, 224, 225, 226, 227, 230, 233, 249, 250, 253, 255, 259*
cardiovascular disease *17, 19*
chronic disease *115, 216*
clinical evidence *47, 49, 50, 51, 52*
clinical research *11, 20, 23, 32, 33, 36, 37, 42, 59, 60, 61, 62, 117, 138, 145, 146, 161, 207, 215, 216, 217, 218, 219, 220, 221, 222, 223, 224, 226*
clinical trials *20, 21, 23, 47, 50, 52, 60, 66, 117, 119, 120, 155, 156, 209, 215, 218, 219, 220, 221, 222, 223, 224, 225, 231, 232*
Commission on Reimbursement of Medicines *42. See also* CRM
cost containment *221, 236*
cost-effectiveness *41, 43, 50, 51, 52, 72, 108, 109*
CRM *42, 43, 48, 49, 176*

D
DALY *19, 233, 234*
developed countries *24, 77, 186, 200*
developing countries *15, 16, 17, 19, 21, 22, 23, 25, 29, 37, 38, 147, 183, 186, 188, 190, 199, 200, 203, 204, 205, 206, 207, 233*
developing world *15, 16, 17, 18, 19, 20, 21, 22, 39, 86, 234*

diabetes *10, 17, 19, 31, 58, 115, 116, 117, 118, 119, 121, 122, 130, 250*
Diabetes Research Center *117, 122. See also* DRC
DRC *117, 120*

E

EBM *42, 43, 44, 45, 50, 52. See also* Evidence Based Medicine
effectiveness *19, 28, 41, 42, 43, 46, 49, 50, 51, 52, 72, 108, 109, 110, 152, 230, 233*
efficiency *15, 41, 79, 90, 92, 99, 100, 101, 102, 107, 111, 132, 152, 174, 187, 188*
EMEA *42, 57, 58, 59, 63, 129, 230. See also* European Medicines Agency
epidemiology *17*
ethical concerns *200, 208*
ethical values *198*
European Commission *30, 31, 45, 48, 120, 169, 207, 212, 213, 220*
European Framework Programmes *31*
European Medicines Agency *129. See also* EMEA
Evidence Based Medicine *42, 145, 152. See also* EBM

F

Framework Convention on Tobacco Control *23*

G

Global Fund *36*
global health *15, 16, 17, 19, 20, 24, 200, 209, 211*
globalization *168*
governance *80, 167, 168, 171, 172, 173, 175, 177, 179*
growth norm *107, 108, 110*

H

Health Care Knowledge Center *131, 140. See also* KCE
health care politics *143*
health economics *15, 195*
health funds *171, 172, 175, 176*
health insurance *29, 33, 82, 99, 102, 103, 105, 106, 107, 111, 112, 125, 171, 172, 173, 175, 176, 179, 180*
health systems *17, 20, 22, 195, 209*
Herceptin *109, 110, 178*
HIV/AIDS *16, 18, 28, 31, 37, 38, 39, 199, 205, 206*
hormone replacement therapy *21, 250, 255*

[263]

I

ICT *90, 91, 94*

inequity *17, 18*

insurance *29, 33, 65, 73, 81, 82, 83, 92, 93, 99, 100, 102, 103, 104, 105, 106, 107, 108, 110, 111, 112, 125, 128, 129, 130, 132, 135, 136, 137, 140, 144, 170, 171, 172, 173, 175, 176, 179, 180, 184, 187, 209*

International Project for Microbicides *37*

Italian Agency for Drugs *59*

K

KCE *108, 158, 162. See also* Health Care Knowledge Center

L

limits *77, 83, 88, 89, 93, 125, 126, 127, 169, 178*

M

malaria *16, 18, 19, 233, 236, 240*

medicalization *255, 256*

medical technology *15*

Millennium Development Goals *16*

molecular biology *217*

molecular therapies *66*

Moureaux Act *172*

N

National Health Service *59, 253. See also* NHS

New Molecular Entities *44. See also* NME

NHS *55, 59, 109, 189, 221. See also* National Health Service

NICE *41, 108, 109, 110, 128*

NIH *19, 22, 30, 158*

NME *32, 44, 45. See also* New Molecular Entities

O

oncology *137, 138, 218, 222, 227*

open source *22, 229, 230, 231, 232, 233, 236, 237, 238, 239*

orphan drugs *10, 41, 42, 44, 46, 47, 48, 49, 50, 51, 52, 54, 55, 59, 60, 177*

orphan medicinal products *55*

P

patents *122, 128, 200, 201, 202, 208, 229, 231, 233, 237, 238*

patient-centered *11, 143, 144, 146, 148, 150, 152*

personalized medicine *10, 65, 66, 68, 70, 72, 74*

pharmaceutical industry *11, 19, 20, 28, 31, 32, 34, 36, 38, 47, 59, 62, 70, 125, 126, 127, 129, 131, 132, 141, 155, 157, 158, 159, 160, 161, 162, 167, 173, 175, 197, 199, 200, 202, 203, 205, 210, 218, 220, 239*

pharmacology *29, 53, 63, 144, 163*

physicians *7, 58, 152, 155, 157, 159, 160, 188, 205, 230*

policy *7, 8, 9, 18, 21, 23, 25, 57, 61, 77, 79, 83, 84, 89, 92, 93, 95, 105, 121, 125, 132, 134, 136, 137, 159, 161, 167, 168, 169, 172, 174, 175, 176, 178, 179, 188, 189, 191, 192, 195, 205, 206, 222*

policy makers *7, 21, 77, 121, 125, 134, 136, 137, 167, 222*

PPPs *22, 36. See also* Public-Private Partnerships

pre-clinical *73, 119*

prescriber *155, 158, 159*

pricing *41, 52, 200, 208, 209, 210, 211*

private expenditure *81, 82, 83, 88*

privatization *107, 186, 187*

public health care expenditure *100*

public-private *183, 184, 186, 189, 190, 191, 192, 230*

Public-Private Partnerships *22, 36, 193. See also* PPPs

R

R&D *3, 19, 22, 31, 32, 39, 44, 48, 118, 126, 128, 129, 133, 139, 140, 201, 229, 230, 233, 234, 235, 237, 238, 239*

registration *16, 23, 32, 36, 46, 65, 70, 73, 157, 158, 178, 209, 218, 235*

reimbursement *29, 30, 31, 41, 42, 43, 44, 45, 46, 47, 48, 49, 50, 51, 52, 55, 59, 65, 73, 91, 128, 157, 158, 175, 176, 177, 178*

S

sacred values *106*

sickness funds *7, 111, 125, 128, 129, 137*

social security *78, 79, 80, 84, 102, 105, 112, 134, 168, 169, 230*

solidarity *10, 29, 93, 99, 101, 102, 103, 105, 106, 107, 110, 127, 133, 134, 135, 176, 187, 193*

stakeholders *7, 8, 24, 94, 121, 125, 127, 130, 132, 137, 139, 140, 155, 167, 175, 222, 223*

T

TB *16, 17, 18*
therapeutic value *34, 41, 42, 43, 44, 45, 46, 48, 49, 50, 51, 52*
toxicities *66*
trade-offs *80, 87, 105, 106, 107, 108, 109, 110, 111, 187*
transparency *8, 23, 24, 42, 90, 92, 105, 106, 107, 108, 111, 132, 140, 148, 151, 152, 156, 157, 159, 160, 171, 172, 175, 238*

U

United Nations *84, 206*

V

Vrije Universiteit Brussel *3, 4, 5, 7, 53, 75, 95, 117, 122, 241*. See also VUB
VUB *95, 102, 117, 120, 122, 229*. See also Vrije Universiteit Brussel
vulgar values *106*

W

World Health Organization *15, 24, 25, 28, 48, 63, 84, 91, 193, 194, 195*
World Medical Association *206*